Lixiong Shao · Jianmei Lu
Min Shi

Chemical Transformations
of Vinylidenecyclopropanes

 Springer

Lixiong Shao
College of Chemistry and Materials
 Engineering
Wenzhou University
Chashan University Town
Wenzhou 325035, Zhejiang
People's Republic of China
e-mail: shaolix@wzu.edu.cn

Jianmei Lu
College of Chemistry and Materials
 Engineering
Wenzhou University
Chashan University Town
Wenzhou 325035, Zhejiang
People's Republic of China
e-mail: ljm@wzu.edu.cn

Min Shi
State Key Laboratory of Organometallic
 Chemistry
Shanghai Institute of Organic
 Chemistry
Chinese Academy of Sciences
345 Lingling Road
Shanghai 200032
People's Republic of China
e-mail: mshi@mail.sioc.ac.cn

ISSN 2191-5407
ISBN 978-3-642-27572-2
DOI 10.1007/978-3-642-27573-9
Springer Heidelberg New York Dordrecht London

e-ISSN 2191-5415
e-ISBN 978-3-642-27573-9

Library of Congress Control Number: 2011945822

Printed on acid-free paper

Springer is part of Springer Science+Business Media (www.springer.com)

Preface

Vinylidenecyclopropanes (VDCPs), which have strained cyclopropyl group connected with an allene moiety and yet are thermally stable and reactive substances in organic chemistry, are versatile intermediates in organic synthesis. During the past decades, VDCPs have demonstrated special reactivities, which can be tuned by the electronic or steric effects and nature of the substituents on the skeleton. Traditionally, great attention and a lot of effects have been focused on the photo- and thermal-induced chemistry of VDCPs. Recently, we have thoroughly investigated the Lewis acid or Brønsted acid, as well as transition metal, catalyzed/mediated chemistry of VDCPs and have found some new reactions of VDCPs, showing their significant usefulness in organic synthesis. In this volume, we will describe our investigations on the chemistry of VDCPs, including their preparation, their reactivities upon treating with Lewis or Brønsted acid, as well as transition metal catalysts and miscellaneous reactions.

Wenzhou, Shanghai, September 2011
Lixiong Shao
Jianmei Lu
Min Shi

Acknowledgments

The authors are deeply grateful to all the co-workers mentioned in the literatures. Financial support from the Shanghai Municipal Committee of Science and Technology (06XD14005 and 08dj1400100-2), National Basic Research Program of China (973)-2009CB825300, the National Natural Science Foundation of China (21072206, 20472096, 20872162, 20672127, 20821002, 20732008 and 21002072) and the Opening Foundation of Zhejiang Provincial Top Key Discipline (No. 100061200123) are greatly acknowledged.

Contents

Chapter 1
Introduction

Vinylidenecyclopropanes (VDCPs) **1** (Fig. 1.1), which have strained cyclopropyl group connected with an allene moiety and yet are thermally stable and reactive substances in organic chemistry, are versatile intermediates in organic synthesis [1, 2]. The first synthesis of VDCPs **1** can be traced back to 1959 [3]. During the past decades, VDCPs **1** have demonstrated special reactivities, which can be tuned by the electronic or steric effects and nature of the substituents on the skeleton. Thermal, photochemical, Lewis or Brønsted acids, as well as transition metals-catalyzed/mediated skeleton conversions of VDCPs **1** have attracted much attention from mechanistic, theoretical, spectroscopic and synthetic viewpoints [4–7]. For a long period of time, VDCPs **1** were considered as highly unstable compounds, and the traditional investigations were focused mainly on the photo- and heat-induced chemistry of VDCPs **1**. However, during the latest years, VDCPs **1** were found to demonstrate good reactivities and selectivities depending on the nature of the electronic and steric effects of the substituents. It is worthy of noting that more reaction patterns were found for VDCPs **1** with aryl substituent(s) during the recent years. For this guidance, this book intends to collect systematically the widespread knowledge not only regarding synthetic methods, but also including advances on the chemistry of VDCPs **1** bearing at least one aryl substituent on the terminal of the allene or cyclopropyl ring moiety, mainly on the Lewis or Brønsted acid-mediated/catalyzed transformations of VDCPs, transition metal-catalyzed transformations of VDCPs and miscellaneous analogs to update the recent review [8]. In order to minimize the overlap with the recent review, similar mechanisms for some of the corresponding reactions will be overlooked in this book. This book will cover the literature up to the early of year 2011.

Generally, VDCPs **1** are prepared through the reaction of alkenes with in situ produced alkenylidenecarbenes [9–12]. These alkenylidenecarbenes can be formed by treating halogenoalkynes [13–26], halogenoallenes [15, 18, 21, 27–31], polyhalogenocyclopropanes [32–41], and polyhalogenoalkanes [42] with strong bases, also can be generated by heating of diazoallenes [43] and so on [44–53].

L. Shao et al., *Chemical Transformations of Vinylidenecyclopropanes*,
SpringerBriefs in Molecular Science, DOI: 10.1007/978-3-642-27573-9_1,
© The Author(s) 2012

Fig. 1.1 Structure of
VDCPs **1**

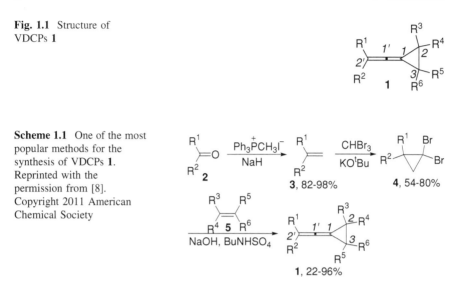

Scheme 1.1 One of the most
popular methods for the
synthesis of VDCPs **1**.
Reprinted with the
permission from [8].
Copyright 2011 American
Chemical Society

The method reported by Mizuno and co-workers in 1991[37] is one of the most popular methods to synthesize various VDCPs, and the general route is shown in Scheme 1.1. The general procedure includes a Wittig reaction of the corresponding carbonyl compounds **2** to form alkenes **3** [54–58]. Then, cyclopropanation of alkenes **3** with the in situ generated dibromocarbene gives the 1,1-dibromocyclopropanes **4** [59]. Finally, under phase-transfer conditions, VDCPs **1** can be obtained in acceptable to high yields via the reaction of 1,1-dibromocyclopropanes **4** with various substituted alkenes **5**.

References

1. Poutsma ML, Ibarbia PA (1971) Electrophilic additions to 2-methyl-1-(tetramethylcyclo-propylidene)propene. Generation of cyclopropylidenecarbinyl cations. J Am Chem Soc 93: 440–450
2. Smadja W (1983) Electrophilic addition to allenic derivatives: chemo-, regio-, and stereochemistry and mechanisms. Chem Rev 83:263–320
3. Hartzler HD (1959) Carbenes from *t*-acetylenic chlorides. Synthesis of alkenylidenecyclo-propanes. J Am Chem Soc 81:2024–2025
4. Pasto DJ, Borchardt JK (1976) Carbon-13 nuclear magnetic resonance spectral properties of alkenylidenecyclopropanes. J Org Chem 41:1061–1063
5. Pasto DJ, Fehlner TP, Schwartz ME, Baney HE (1976) On the orbital interactions of three-membered rings with π systems. Electronic structure of alkenylidenecyclopropanes. J Am Chem Soc 98:530–534
6. Chapman OL, Gano J, West PR (1981) Acenaphthyne. J Am Chem Soc 103:7033–7036
7. Zhanpeisov NU, Mizuno K, Anpo M, Leszczynski J (2004) C1–C2 bond cleavage in vinylidenecyclopropanes: theoretical density functional theory study. Int J Quant Chem 96:343–348

8. Shi M, Shao LX, Lu JM, Wei Y, Mizuno K, Maeda H (2010) Chemistry of vinylidenecyclopropanes. Chem Rev 110:5883–5913
9. Stang PJ (1978) Unsaturated carbenes. Chem Rev 78:383–405
10. Stang PJ (1982) Recent developments in unsaturated carbenes and related chemistry. Acc Chem Res 15:348–354
11. Stang PJ, Ladika M (1981) Extended unsaturated carbenes. Generation, nature, and chemistry of alkatetraenylidenecarbenes, $R_2C=C=C=C=C=C$. J Am Chem Soc 103:6437–6443
12. Hoffmann RW, Riemann A, Mayer B (1985) Carben-reaktionen, XVIII. Themisches verhalten von 7-alkylidenbicyclo[2.2.1]heptadien-derivaten. Chem Ber 118:2493–2513
13. Eguchi S, Ikemoto T, Kobayakawa Y, Sasaki T (1985) Mild generation of adamantylidenevinylidene (an alkenylidenecarbene) from 2-bromo-2-(trimethylsilylethynyl) adamantane and (2-bromo-2-trimethylsilylvinylidene)adamantine. J Chem Soc Chem Commun 958–959
14. Sheu JH, Yen CF, Huang CW (1993) The regioselectivity of (dimethylvinylidene)carbene directed by aromatic ring upon the addition into aryl alkenes in the presence of other olefins. J Chin Chem Soc 40:59–65
15. Katsuhira T, Harada T, Oku A (1994) New method for generation of alkenylidenecarbenes from propargylic methanesulfonates and its use in regioselective C–H insertion reactions. J Org Chem 59:4010–4014
16. Sheu JH, Yen CF, Chan YL, Chung JF (1990) Regioselective cyclopropanation of (dimethylvinylidene)carbene into the α, β-unsaturated double bond of allylic alcohols in the presence of other olefinic groups. J Org Chem 55:5232–5233
17. Sasaki T, Eguchi S, Ogawa T (1974) Reactions of isoprenoids. XIX. Phase-transfer catalyzed synthesis of dimethylvinylidenecyclopropane derivatives in aqueous medium. J Org Chem 39:1927–1930
18. Hartzler HD (1964) Vinylidene carbenes by α-elimination from haloallenes. J Org Chem 29:1311–1312
19. Hennion GF, Motier JF (1969) Substituted acetylenes. LXXXVII. Solvent effects and nucleophile competition in reactions of 3-chloro-3-methyl-1-butyne. J Org Chem 34:1319–1323
20. Eguchi S, Arasaki M (1988) Synthesis of novel carbo- and heteropolycycles. Part 8. An efficient and convenient synthesis of adamantylidenevinylidenecyclopropane derivatives. J Chem Soc Perkin Trans 1:1047–1050
21. Sheu JH, Yen CF, Huang CW, Chan YL (1991) Further study on the regioselectivity of (dimethylvinylidene)-carbene uopn the cycloaddition to the α, β-unsaturated double bond of an allylic alcohol and the normal olefinic group. Tetrahedron Lett 32:5547–5550
22. Sasaki T, Eguchi S, Ohno M, Nakata F (1976) Crown ether catalyzed synthesis of dialkylvinylidenecyclopropane derivatives. J Org Chem 41:2408–2411
23. Hartzler HD (1961) Carbenes from derivatives of ethynylcarbinols. The synthesis of alkenylidenecyclopropanes. J Am Chem Soc 83:4990–4996
24. Liese T, de Meijere A (1986) Vielseitige synthese von alkinylcyclopropanen aus olefin-perchlorvinylcarben-addukten. Chem Ber 119:2995–3026
25. Crombie L, Maddocks PJ, Pattenden G (1978) Monoterpene synthesis via alkenylidene cyclopropanes: acid- and base-catalysed rearrangements. Tetrahedron Lett 19:3479–3482
26. Maercker A, Wunderlich H, Girreser U (1996) Polylithiumorganic compounds-23. 3,4-Dilithio-1,2-butadienes by addition of lithium metal to 1,4-unsymmetrically substituted butatrienes. Tetrahedron 52:6149–6172
27. Landor SR, Whiter PF (1965) Allenes. IX. Carbenes from 1-haloallenes. J Chem Soc 5625–5629
28. Patrick TB, Haynie EC, Probst WJ (1972) A comparative study of some reactions of dimethylvinylidene and dimethylmethylidene. J Org Chem 37:1553–1556
29. Patrick TB, Schmidt DJ (1977) Concerning the nature of dimethylvinylidenecarbene. J Org Chem 42:3354–3356

30. Landor SR, Patel AN, Whiter PF, Greaves PM (1966) Allenes. XI. The preparation of 3-alkyl- and 3,3-dialkyl-1-bromoallenes. J Chem Soc C 1223–1226
31. Aue DH, Meshishnek MJ (1977) Synthesis and thermal rearrangement of 3-(2'-methyl-prop-1'-enylidene)tricyclo[3.2.1.0^{2,4}]oct-6-ene. J Am Chem Soc 99:223–231
32. Sugita H, Mizuno K, Mori T, Isagawa K, Otsuji Y (1991) Unusual mode of addition of 1,2-alkadienylidene carbenes to 1,3-dienes: 1,4-addition to rigid and flexible 1,3-dienes. Angew Chem Int Ed Engl 30:984–986
33. Al-Dulayymi J, Baird MS (1988) Highly functionalised carbenes and cyclopropenes from tetrahalocyclopropanes. Tetrahedron Lett 29:6147–6148
34. Le Perchec P, Conia JM (1970) Etude des rotanes(III) le bicyclopropylidène et sa dimérisation thermique en tétracyclopropylidène. Tetrahedron Lett 11:1587–1588
35. Denis JM, Le Perchec P, Conia JM (1977) Etude des petits cycles—XXXVII: syntheses et proprietes spectrales du tetracyclopropylidene ([4]rotane) et des composes polycyclopropylspiraniques de la serie. Tetrahedron 33:399–408
36. Lukin KA, Kozhushkov SI, Andrievsky AA, Ugrak BI, Zefirov NS (1991) Synthesis of branched triangulanes. J Org Chem 56:6176–6179
37. Isagawa K, Mizuno K, Sugita H, Otsuji Y (1991) J Chem Soc Perkin Trans 1 2283–2285
38. Zöllner S, Buchholz H, Boese R, Gleiter R, de Meijere A (1991) 7,7'-Bi(dispiro[2.0.2.1] heptylidene)—the perspirocyclopropanated bicyclopropylidene. Angew Chem Int Ed Engl 30:1518–1520
39. Averina EB, Karimov RR, Sedenkova KN, Grishin YK, Kuznetzova TS, Zefirov NS (2006) Carbenoid rearrangement of *gem*-dihalogenospiropentanes. Tetrahedron 62:8814–8821
40. Lukin KA, Zefirov NS, Yufit DS, Struchkov YT (1992) Unusual rearrangement of triangulane *gem*-dibromides in the presence of methyllithium. Tetrahedron 48:9977–9984
41. Billups WE, Haley MM, Boese R, Bläser D (1994) Synthesis of the bicyclopropenyls. Tetrahedron 50:10693–10700
42. Keyaniyan S, Gäthling W, de Meijere A (1984) Convenient syntheses of dichloroethenylidenecyclopropanes: precursors to difunctional cyclopropane derivatives. Tetrahedron Lett 25:4105–4108
43. Northington DJ, Jones WM (1971) Diphenyldiazoallene and diphenylallenyl diazotate. Tetrahedron Lett 12:317–320
44. Patrick TB (1974) Phase-transfer catalyzed generation of dimethylvinylidene carbene. Tetrahedron Lett 15:1407–1408
45. Bleiholder F, Shechter H (1964) Substituted dimethylenecyclopropanes. Capture reactions of alkylidene and vinylidene carbenes by allenes. J Am Chem Soc 86:5032–5033
46. Tsuno T, Sugiyama K (1995) Allenyl(vinyl)methane photochemistry. Photochemistry of 5-[2-(1,2-propadienyl)-substituted alkylidene]-2,2-dimethyl-1,3-dioxane-4,6-diones. Bull Chem Soc Jpn 68:3175–3188
47. Tsuno T, Sugiyama K (1999) Allenyl(vinyl)methane photochemistry. Photochemistry of methyl 4,4-dimethyl-2,5,6-heptatrienoate derivatives. Bull Chem Soc Jpn 72:519–531
48. Tsuno T, Sugiyama K (1991) Photochemistry of isopropylidene 3,3,6-trimethyl-1,4,5-heptatriene-1,1-dicarboxylate and its homologues. Chem Lett 20:503–506
49. Stang PJ, Fisk TE (1979) Extented unsaturated carbenes. Generation and reaction of allenylidene carbenes (R)2C=C=C=C: with olefins. J Am Chem Soc 101:4772–4773
50. Stang PJ, Fisk TE (1980) Extented unsaturated carbenes. Generation and nature of alkadienylidenecarbenes. J Am Chem Soc 102:6813–6816
51. de Meijere A, Jaekel F, Simon A, Borrmann H, Köhler J, Johnels D, Scott LT (1991) Regioselective coupling of ethynylcyclopropanes units: hexaspiro[2.0.2.4.2.0.2.4.2.0.2.4] triaconta-7,9,17,19,27,29-hexayne. J Am Chem Soc 113:3935–3941
52. Manhart S, Schier A, Paul M, Riede J, Schmidbaur H (1995) New organophosphorus ligands: cyclopropanation and other reactions of cumulenes bearing diphenylphosphanyl substituents. Chem Ber 128:365–371

53. Campbell MJ, Pohlhaus PD, Min G, Ohmatsu K, Johnson JS (2008) An "anti-Baldwin" 3-*exo-dig* cyclization: preparation of vinylidene cyclopropanes from electron-poor alkenes. J Am Chem Soc 130:9180–9181

54. Wittig G, Schöllkopf U (1954) Über triphenyl-phosphin-methylene als olefinbildende reagenzien I. Mitteil Chem Ber 97:1318–1330

55. Edmonds M, Abell A (2004) Modern carbonyl olefination. In: Takeda T (ed) Wiley, Weinheim, pp 1–17

56. Lawrence NJ (1996) Preparation of alkenes. In: Williams JMJ (ed) Oxford University Press, Oxford, pp 19–58

57. Murphy PJ, Brennan J (1988) The Wittig olefination reaction with carbonyl compounds other than aldehydes and ketones. Chem Soc Rev 17:1–30

58. Maryanoff BE, Reitz AB (1989) The Wittig olefination reaction and modifications involving phosphoryl-stabilized carbanions. Stereochemistry, mechanism, and selected synthetic aspects. Chem Rev 89:863–927

59. Sandler SR (1977) Chain elongation of alkenes via gem-dihalocyclopropanes: 2-bromo-3, 3-diphenyl-2-propen-1-yl acetate. Org Synth 56:32–35

Chapter 2
Lewis or Brønsted Acid-Mediated Transformations of VDCPs

Abstract Lewis or Brønsted acid-mediated intramolecular rearrangement of VDCPs, and the reactions of VDCPs with acetals, aldehydes, ketones, imines, activated alkenes, nitriles, acyl chlorides, and alcohols are described in this chapter.

Keywords Vinylidenecyclopropanes · Intramolecular rearrangement · Cycloaddition reaction · Ring-opening reaction · Mercury acetate-mediated reaction · Lewis acid-mediated reaction · Brønsted acid-mediated reaction

2.1 Mercury Acetate-Mediated Transformations of VDCPs

Pasto et al. reported the mercury acetate-mediated transformations of VDCPs **1** in 1976 [1–5]. Acetoxymercuration of VDCP **1a** followed by reductive demercuration using a great excess of sodium borohydride produced a complex mixture of the monomeric acetates **8** and **9** (60:40 ratio), dimeric diacetates, and bis(acetoxyalkyl)mercury compounds. Disrotatory ring-opening of an intermediate spiromercurinium ion **6** (or possible cyclopropyl cation **7** as a transition state) is expected to occur with outward rotation of the phenyl group, i.e., in the least sterically congested manner, to produce an allylic cation which then reacts with acetate to produce products **8** and **9**. In addition, **9** can be cleanly rearranged to **8** in the presence of strong protic acid (Scheme 2.1).

In addition to the formation of alcohols **10** and **11**, which correspond to the acetates **8** and **9** formed in acetoxymercuration of VDCP **1a**, hydroxymercuration of **1a** in 50% aqueous tetrahydrofuran (THF) also resulted in the formation of the acetylenic alcohol **12** (small amounts of acetates **8**, **9**, and **13** are also formed) (Scheme 2.2).

L. Shao et al., *Chemical Transformations of Vinylidenecyclopropanes*,
SpringerBriefs in Molecular Science, DOI: 10.1007/978-3-642-27573-9_2,
© The Author(s) 2012

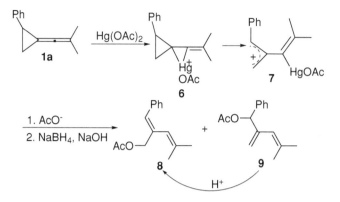

Scheme 2.1 Hg(OAc)$_2$-mediated transformation of VDCP **1a**. Reprinted with the permission from Ref. [1]. Copyright 2011 American Chemical Society

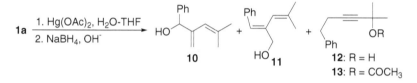

Scheme 2.2 Hg(OAc)$_2$-mediated reaction of VDCP **1a** in aqueous THF. Reprinted with the permission from Ref. [1]. Copyright 2011 American Chemical Society

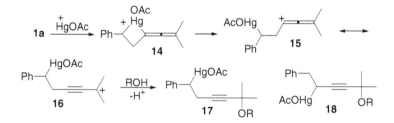

Scheme 2.3 Plausible mechanism for the Hg(OAc)$_2$-mediated reaction of VDCP **1a**. Reprinted with the permission from Ref. [1]. Copyright 2011 American Chemical Society

The alcohol **12** and acetate **13** may be formed by initial attack by acetoxy-mercury cation on one of the ring bonds, either as shown to produce **17** or alternatively on the –CH$_2$–C= bond to give **18**, both of which would be reduced to **12** and **13** (Scheme 2.3) [6, 7].

Scheme 2.4 Sn(OTf)$_2$-catalyzed intramolecular rearrangement of VDCPs **1**. Reprinted with the permission from Ref. [1]. Copyright 2011 American Chemical Society

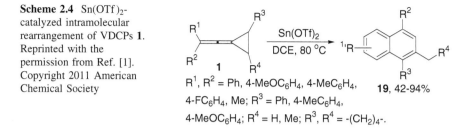

R^1, R^2 = Ph, 4-MeOC$_6$H$_4$, 4-MeC$_6$H$_4$, 4-FC$_6$H$_4$, Me; R^3 = Ph, 4-MeC$_6$H$_4$, 4-MeOC$_6$H$_4$; R^4 = H, Me; R^3, R^4 = -(CH$_2$)$_4$-.

19, 42–94%

2.2 Lewis or Brønsted Acid-Mediated Intramolecular Rearrangement of VDCPs

During the recent years, Shi et al. have reported a series of Lewis acid-catalyzed rearrangement reactions of VDCPs **1**. For example, in 2005, Shi et al. first found the Lewis acid-catalyzed rearrangement reaction of VDCPs **1**. It was observed that VDCPs **1** could rearrange to naphthalene derivatives **19** in the presence of Sn(OTf)$_2$, in acceptable to high yields under mild conditions (Scheme 2.4) [1, 8].

Based on this pioneering work, Shi et al. investigated thoroughly the Lewis acid-catalyzed rearrangement reactions of VDCPs **1** having three substituents on the corresponding cyclopropyl rings [9–11]. It was found that the reaction products are highly dependent on the substituents on the corresponding cyclopropyl rings and the electronic nature of the aryl groups on VDCPs **1**. For VDCPs **1** bearing two alkyl groups at the C3 position (R^1, R^2, R^3 = aryl; R^4 = H; R^5, R^6 = alkyl), naphthalene derivatives **19** were formed in the presence of Lewis acid Eu(OTf)$_3$ in DCE at 40 °C. For VDCPs **1** in which R^1, R^2, R^3 = aryl and R^4, R^5 = alkyl (*syn/anti* isomeric mixture), the corresponding 6aH-benzo[*c*]fluorine derivatives **20** were obtained in the *syn*-configuration via a double intramolecular Friedel–Crafts reaction using the substrates without electron–withdrawing substituents on aryl groups or the corresponding indene derivatives **21** were formed via an intramolecular Friedel–Crafts reaction as long as one electron-deficient aryl group was attached. For VDCPs **1** in which R^1, R^2, R^3, R^4 = aryl and R^5 = alkyl or H, the corresponding indene derivatives **21** were obtained exclusively via a sterically demanding intramolecular Friedel–Crafts reaction (Scheme 2.5).

Plausible mechanisms for the formation of naphthalene, 6aH-benzo[*c*]fluorine, and indene derivatives are shown as below: the coordination of VDCPs **1** to Lewis acid initially gives 1-cyclopropylvinyl cation **22**, a vinyl group stabilized cyclopropyl cation [12–16], which results in the formation of cyclopropyl ring-opened cationic intermediate **23** or its resonance-stabilized zwitterionic intermediate **24** and **24'**, which is stabilized by the aromatic R^3 group in most cases. The intramolecular Friedel–Crafts reaction [17] produces the cyclized intermediate **25**, from which the thermodynamically favored naphthalene derivatives **19** are formed via successive 1,3-carbocation rearrangement, 1,4-proton shift along with release of Lewis acid or deprotonation and 1,3-proton shift (Scheme 2.6, **path a**) [18].

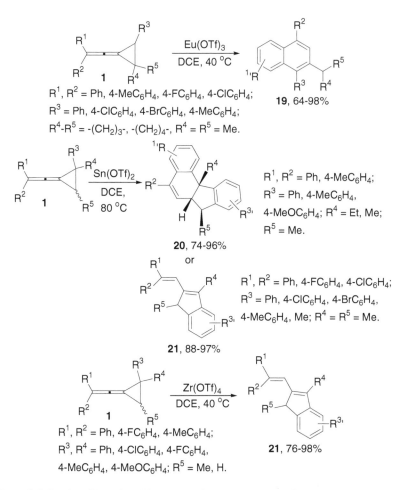

Scheme 2.5 Lewis acid-catalyzed intramolecular rearrangement of VDCPs **1**. Reprinted with the permission from Ref. [1]. Copyright 2011 American Chemical Society

6aH-benzo[c]fluorine derivatives **20** can be obtained via a double Friedel–Crafts reaction as shown in **path b** in Scheme 2.6. The formation of indene derivatives **21** is illustrated in **path c** in Scheme 2.6.

Using VDCPs **1** with tetramethyl-substituted cyclopropyl ring ($R^3 = R^4 = R^5 = R^6 = Me$), interesting intramolecular rearrangement patterns were found in the presence of different Lewis or Brønsted acid catalyst under different reaction temperature. For example, with soft Lewis acid Au(I) as the catalyst, 2-isopropylidene-1,1-dimethyl-1,2-dihydronaphthalene derivatives **30** and/or **30'** were obtained in good to high yields at room temperature (Scheme 2.7) [19], while with hard Brønsted acid such as Tf$_2$NH (Tf = trifluoromethanesulfonyl) or toluene-4-sulfonic acid (p-TSA) as the catalyst, the corresponding naphthalene

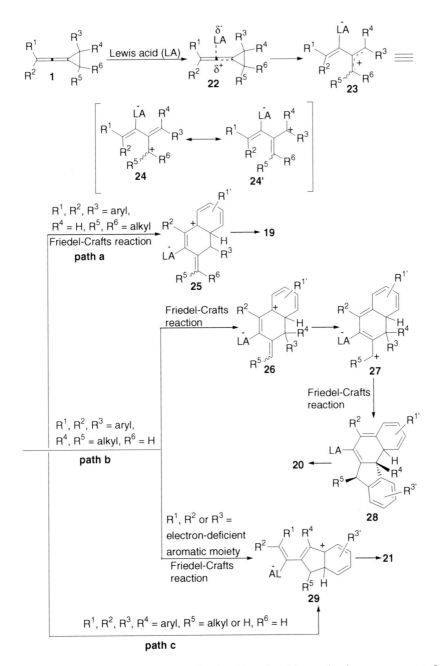

Scheme 2.6 Plausible mechanism for the Lewis acid-catalyzed intramolecular rearrangement of VDCPs **1**. Reprinted with the permission from Ref. [1]. Copyright 2011 American Chemical Society

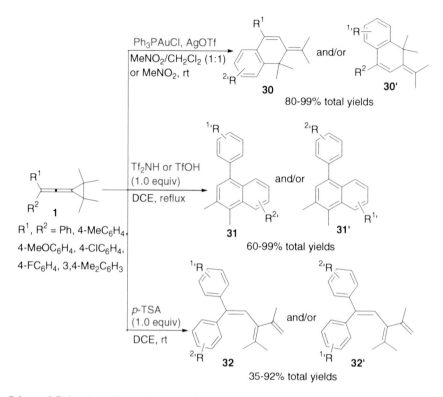

Scheme 2.7 Lewis or Brønsted acid-catalyzed intramolecular rearrangement of VDCPs **1** with tetramethyl groups on the cyclopropyl ring

derivatives **31** and/or **31′** were formed in good to high yields upon heating, along with release of a propene molecule. In addition, in the presence of Brønsted acid such as toluene-4-sulfonic acid (*p*-TSA), the corresponding triene derivatives **32** and **32′** were afforded in moderate to good yields at room temperature (Scheme 2.7) [20]. All ratios of **30** and **30′**, **31** and **31′**, and **32** and **32′** were dependent on the electron character of groups R^1 and R^2.

Plausible mechanism for the formation of products **30** with VDCP **1b** as the model is shown below: first, the corresponding cyclopropyl ring-opened zwitterionic intermediate **34** or the resonance-stabilized zwitterionic intermediate **34′** is formed from the initial zwitterionic intermediate **33**. Then intramolecular Friedel–Crafts reaction with the adjacent aromatic ring takes place to produce zwitterionic intermediate **35**, which affords the corresponding intermediate **36** via deprotonation. Subsequent protonation of intermediate **36** produces the corresponding product **30a** along with the release of Au(I) catalyst (Scheme 2.8).

Plausible mechanism for the formation of products **31** and **32** is shown below: first, protonation of VDCP **1b** by the Brønsted acid regioselectively gives intermediate **37**, which results in the formation of cyclopropyl ring-opened cationic

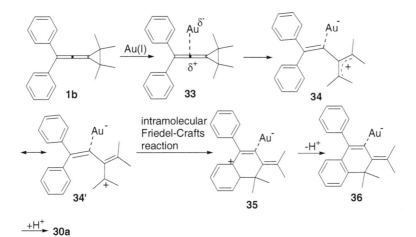

Scheme 2.8 Plausible mechanism for the Au(I)-catalyzed intramolecular rearrangement of VDCP **1b**. Reprinted with the permission from Ref. [19]. Copyright 2011 American Chemical Society

intermediate **38** or **38′**. Then product **32a** will be achieved with release of a proton (Scheme 2.9, **path a**). Alternatively, intermediate **38′** will be transformed to intermediate **39** through an intramolecular Friedel–Crafts reaction (Scheme 2.9, **path b**). Intermediate **39** undergoes subsequent protonation to give intermediate **40**. 1,2-Migration of the methyl group in intermediate **40** gives intermediate **41**, which will produce the final product **31a** with release of a proton and a propene. It is worth noting that this mechanism is postulated on the DFT studies performed with the Gaussian03 program by using the B3LYP method [20].

In the early 2008, Huang et al. reported a Lewis acid TiCl$_4$-mediated ring-expansion reaction of bicyclic VDCPs **1** for the formation of medium- and large-size naphthalenacarbocycle derivatives **42** (Scheme 2.10) [21].

2.3 Lewis or Brønsted Acid-Mediated Reactions of VDCPs with Acetals, Aldehydes and Ketones

Lewis acid Sc(OTf)$_3$-catalyzed reactions of VDCPs **1** with acetals **43** were also established for the preparation of indene derivatives **44** and **45**. This reaction is believed to proceed via regioselective addition of oxonium intermediate to VDCPs **1** and the subsequent intramolecular Friedel–Crafts reaction. It was found that the electronic nature of substituent strongly influenced the reaction results, even leading to different products (Scheme 2.11) [1, 22]. In a word, when R^3, R^4 = aryl, R^5, R^6 = Me or H, indene derivatives **44** can be formed singly; and while R^3 = R^4 = R^5 = R^6 = Me, indene derivatives **45** were obtained exclusively.

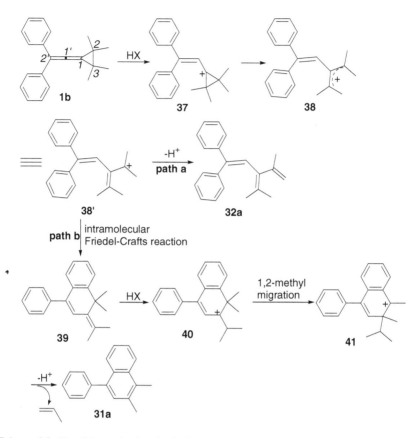

Scheme 2.9 Plausible mechanism for the Brønsted acid-catalyzed intramolecular rearrangement of VDCP **1b**

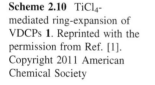 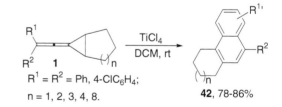

Scheme 2.10 TiCl$_4$-mediated ring-expansion of VDCPs **1**. Reprinted with the permission from Ref. [1]. Copyright 2011 American Chemical Society

$R^1 = R^2 = Ph, 4-ClC_6H_4;$
$n = 1, 2, 3, 4, 8.$

42, 78–86%

Interestingly, the substrate VDCP **1c** with only one phenyl group at the cyclopropyl ring demonstrated a special reactivity under identical reaction conditions. A new product 6-methyl-5,7-diphenyl-7H-benzo[c]fluorine **46a** along with other unidentified by-products was obtained although the yield (16%) was rather low (Scheme 2.12).

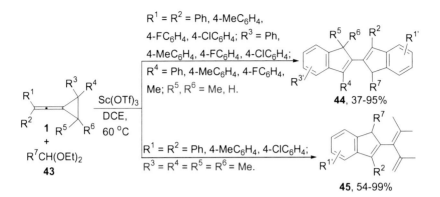

Scheme 2.11 Sc(OTf)$_3$-catalyzed reactions of VDCPs **1** with acetals. Reprinted with the permission from Ref. [1]. Copyright 2011 American Chemical Society

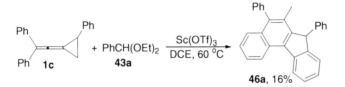

Scheme 2.12 Sc(OTf)$_3$-catalyzed reaction of VDCP **1c** with acetal **43a**. Reprinted with the permission from Ref. [1]. Copyright 2011 American Chemical Society

Huang et al. developed the reactions of VDCPs **1** with aldehydes to provide an efficient and selective method for the synthesis of benzo[c]fluorene, tetrahydro-furan, furan, and indene derivatives. A variety of benzo[c]fluorene **47**, furan **48**, tetrahydrofuran **49**, and indene derivatives **50** can be obtained selectively depending on the Lewis acid catalysts, the substituents on VDCPs **1**, and the reaction temperature [23–25]. For instance, using FeCl$_3$/TMSCl as the catalyst at -10 °C, benzo[c]fluorene derivatives **47** can be obtained in moderate to good yields (R^1, R^2 = aryl; R^3 = aryl, alkyl; R^4 = R^6 = H, R^5 = H, Me); first treat-ment with BF$_3$·OEt$_2$ at -10 °C, then with TMSCl at room temperature, furan derivatives **48** can be achieved in acceptable to moderate yields [R^1 = R^2 = aryl; R^4 = R^6 = H, R^3, R^5 = –(CH$_2$)$_n$– (n = 3, 4, 6, 10)]; using BF$_3$·OEt$_2$ as the catalyst at -10 °C, tetrahydrofuran derivatives **49** can be obtained in acceptable to high yields [R^1 = aryl; R^2 = aryl, alkyl; R^3 = R^4 = R^5 = R^6 = Me or R^3 = R^5 = H, R^4, R^6 = –(CH$_2$)$_n$– (n = 3, 4, 6, 10)]; while using BF$_3$·OEt$_2$ as the catalyst at room temperature, indene derivatives **50** can be formed in low to moderate yields (R^1, R^2 = aryl; R^3 = R^4 = R^5 = R^6 = Me) (Scheme 2.13).

Plausible mechanism for the formation of products **47–50** is outlined below. Initially, the coordination of aldehydes to Lewis acid (LA) gives intermediate **51**, which adds to VDCPs **1** regioselectively to produce intermediate **52**. Then the

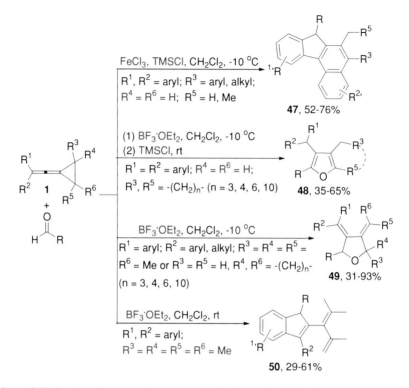

Scheme 2.13 Lewis acid-catalyzed reactions of VDCPs **1** with aldehydes

cyclopropyl ring-opened intermediate **53** or the resonance-stabilized intermediates **53′** and **53″** will be formed. Intramolecular *O*-attack cyclization of intermediate **53′** will afford products **49**. When $R^4 = R^6 = H$, aromatization of **49** will take place to give products **48** (Scheme 2.14, **path a**). When $R^3 = R^4 = R^5 = R^6 = Me$, deprotonation of intermediate **53″** will form intermediate **54**, which then transforms to intermediate **55**. The subsequent intramolecular Friedel–Crafts reaction of intermediate **55** furnishes the indene derivatives **50** (Scheme 2.14, **path b**). In addition, when $R^4 = R^6 = H$, intermediate **53′** undergoes intramolecular Friedel–Crafts reaction to give intermediate **56**. Aromatization of **56** produces the thermodynamically favored naphthalene intermediate **57**, which then transforms to intermediate **58**. Finally, intermediate **58** undergoes another intramolecular Friedel–Crafts reaction to give products **47** (Scheme 2.14, **path c**).

Further studies revealed that using VDCPs **1** with functionalized groups on the cyclopropyl ring shown below as the substrates, their reactions with aldehydes will afford the furo[2,3-*b*]furan derivatives **59** in 37–64% yields (Scheme 2.15).

Plausible mechanism for the formation of products **59** is depicted in Scheme 2.16. First, the addition of the Lewis acid-activating aldehydes **51** to VDCPs **1** may occur to give the ring-opened intermediate **60** via proximal cleavage of the cyclopropyl ring. Then an intramolecular cyclization of

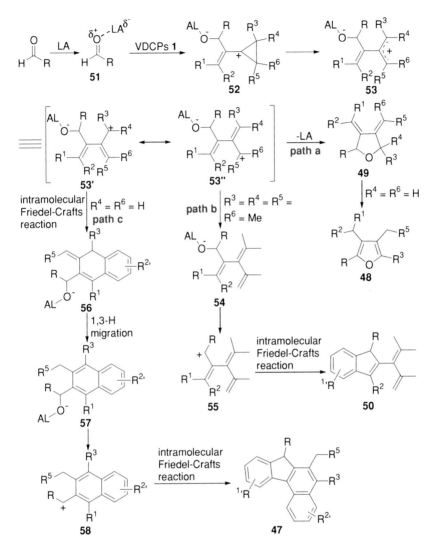

Scheme 2.14 Plausible mechanism for the Lewis acid-catalyzed reactions of VDCPs **1** with aldehydes. Reprinted with the permission from Ref. [25]. Copyright 2011 American Chemical Society

intermediate **60** along with the release of the Lewis acid furnished the [3 + 2] cycloaddition intermediate **61**, which will transform to the final products **59** via 1,3-H shift.

The reactions of VDCPs **1** with activated carbonyl compounds **62** were also investigated and it was found that a number of functionalized tetrahydrofurans **63** and 3,6-dihydropyrans **64** can be formed in moderate to good yields selectively in the presence of different Lewis acids (Scheme 2.17) [26]. In these reactions,

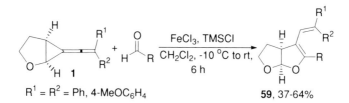

R^1 = R^2 = Ph, 4-MeOC$_6$H$_4$ **59**, 37-64%

Scheme 2.15 Lewis acid-catalyzed reactions of bicyclic VDCPs **1** with aldehydes

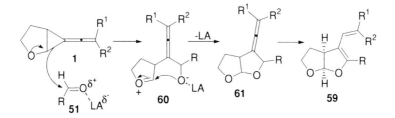

Scheme 2.16 Plausible mechanism for the Lewis acid-catalyzed reactions of bicyclic VDCPs **1** with aldehydes. Reprinted with the permission from Ref. [24]. Copyright 2011 Wiley John and Sons

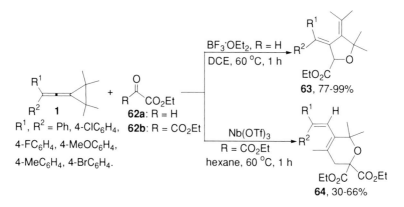

Scheme 2.17 Lewis acid-catalyzed reactions of VDCPs **1** with activated carbonyl compounds. Reprinted with the permission from Ref. [1]. Copyright 2011 American Chemical Society

tetrahydrofurans **63** were obtained in 77–99% yields in the reactions of VDCPs **1** with oxo-acetic acid ethyl ester **62a** in the presence of BF$_3$·OEt$_2$; however, 3,6-dihydropyrans **64** were formed in 30-66% yields in the reactions of VDCPs **1** with 2-oxo-malonic acid diethyl ester **62b** in the presence of Nb(OTf)$_3$.

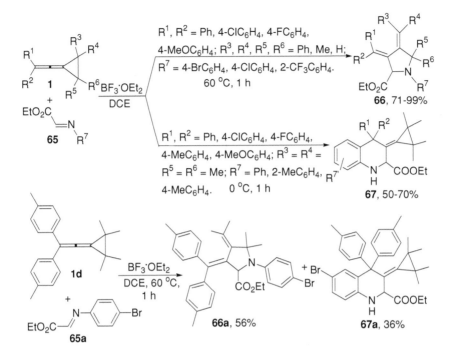

Scheme 2.18 BF$_3$·OEt$_2$-catalyzed reactions of VDCPs **1** with ethyl (arylimino)acetates. Reprinted with the permission from Ref. [1]. Copyright 2011 American Chemical Society

2.4 Lewis or Brønsted Acid-Mediated Reactions of VDCPs with Imines and Their Analogs

The scope of Lewis acid-catalyzed reaction of VDCPs **1** was further extended with respect to a series of ethyl (arylimino)acetates **65**, which led to a facile synthetic protocol of pyrrolidine and 1,2,3,4-tetrahydroquinoline derivatives. A number of pyrrolidine derivatives **66** and 1,2,3,4-tetrahydroquinoline derivatives **67** can be obtained selectively in moderate to good yields by the reaction of VDCPs **1** with ethyl (arylimino)acetates **65** in the presence of Lewis acid BF$_3$·OEt$_2$ depending on the electronic nature of both **65** and R^1 or R^2 aromatic groups of **1** [1, 27–29]. Generally, when R^7 group on **65** is an electron-poor aromatic group, the pyrrolidines **66** will be formed solely; when R^7 is an electron-rich aromatic group, the 1,2,3,4-tetrahydroquinolines **67** will be obtained as the sole products. Meanwhile, if R^1 and R^2 are both electron-rich aromatic groups (R^1 = R^2 = 4-MeC$_6$H$_4$ as the example in this case), both of the products **66a** and **67a** can be obtained in spite of R^7 is an electron-poor or -rich group (Scheme 2.18).

Plausible mechanism for the formation of pyrrolidines **66** and 1,2,3,4-tetrahydroquinolines **67** is outlined in Scheme 2.19. First, ethyl (arylimino)acetate **65** is activated by BF$_3$·OEt$_2$ to afford intermediate **68**, which subsequently adds to C1′

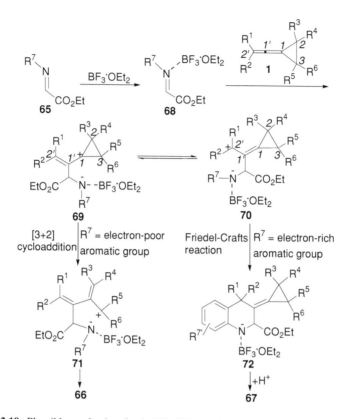

Scheme 2.19 Plausible mechanism for the BF$_3$·OEt$_2$-catalyzed reactions of VDCPs **1** with ethyl (arylimino)acetates. Reprinted with the permission from Ref. [1]. Copyright 2011 American Chemical Society

position of VDCPs **1** to give the corresponding allylic carbocationic intermediates **69** and **70** [30–32]. Intermediate **71**, derived from **69** via a cyclopropyl ring-opening process, undergoes cyclization to give the corresponding [3 + 2] cyclo-addition products **66** when R^7 is an electron-poor aromatic group. Alternatively, if R^7 is an electron-rich aromatic group, intramolecular Friedel–Crafts reaction takes place from intermediate **70** to give intermediate **72** [33], which finally furnishes products **67** [34–37].

During the ongoing investigations, Shi et al. found that VDCPs **1** can also react with other activated carbon–nitrogen, nitrogen–nitrogen, and iodine-nitrogen double-bond-containing compounds, such as N-toluene-4-sulfonyl (N-Ts) imines, diisopropylazodicarboxylate (DIAD), N-tert-butoxycarbonyl (N-Boc) aldimine, and N-Ts-iminophenyliodinane in the presence of Lewis or Brønsted acid to afford the corresponding cycloadducts in moderate to high yields (Scheme 2.20) [38].

Plausible mechanism for the formation of products **76** and **77** is illustrated in Scheme 2.21. Initially, N-Boc aldimine is activated by Brønsted acid TfOH to give

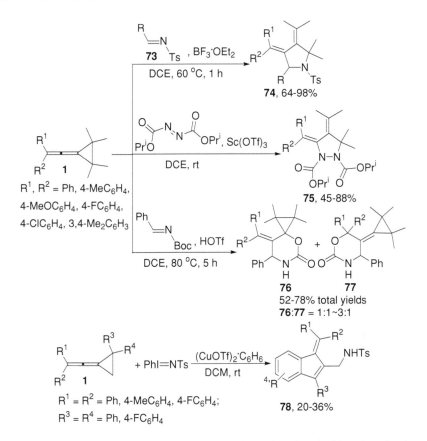

Scheme 2.20 Lewis or Brønsted acid-catalyzed reactions of VDCPs **1** with activated carbon–nitrogen, nitrogen–nitrogen, and iodine-nitrogen double-bond-containing compounds

intermediate **79**, which subsequently adds to the C1′ position of VDCPs **1** selectively to afford the corresponding intermediates **80** and **80′**. Then, ring-closure of intermediates **80** and **80′** affords intermediates **81** and **82**, respectively. Finally, hydrolysis of intermediates **81** and **82** furnishes products **76** and **77**, respectively. It was believed that cyclization of intermediates **80** and **80′** is faster than the cyclopropyl ring-opening reaction in this particular case, so the six-membered products **76** and **77** were formed instead of the [3 + 2] cycloaddition products.

Plausible mechanism for the formation of products **78** is outlined below: initially, intermediate **83** is formed by the reaction of Cu(I) with PhI = NTs according to the general mechanism in the Cu(I)-catalyzed aziridination of olefin [39]. Then, the reaction of intermediate **83** with VDCPs **1** gives the ring-opened zwitterionic intermediate **84** presumably through nucleophilic attack of the cyclopropyl ring on the nitrene species. The allylic rearrangement of zwitterionic intermediate **84** produces intermediate **85**, which undergoes an intramolecular

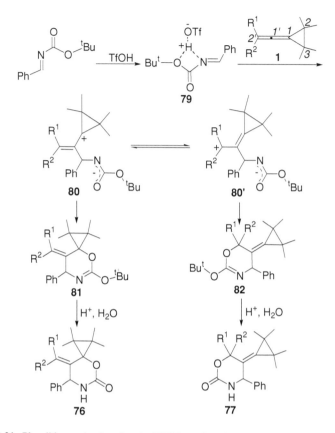

Scheme 2.21 Plausible mechanism for the TfOH-catalyzed reactions of VDCPs **1** with *N*-Boc aldimine

Friedel–Crafts reaction with the R^4 group to generate intermediate **86**. Protonation of intermediate **86** will furnish products **78** (Scheme 2.22).

2.5 Brønsted Acid-Mediated Reactions of VDCPs with Activated Alkenes

Further studies showed that VDCPs **1** can also react with activated olefins such as α,β-unsaturated ketones (aldehyde) **87** to give the corresponding [3 + 2] cycloaddition products **88** as the major ones in moderate to high yields (Scheme 2.23) [40].

The formation of products **88** can be rationalized as below with the cycloaddition reaction of VDCP **1b** and methyl vinyl ketone (MVK) as the model: first, MVK is protonated by Tf_2NH to generate intermediate **89a**, which undergoes

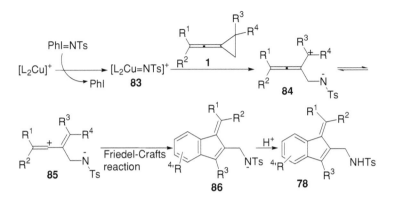

Scheme 2.22 Plausible mechanism for the Cu(I)-catalyzed reactions of VDCPs **1** with *N*-Ts-iminophenyliodinane

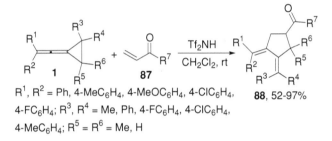

R^1, R^2 = Ph, 4-MeC$_6$H$_4$, 4-MeOC$_6$H$_4$, 4-ClC$_6$H$_4$,
4-FC$_6$H$_4$; R^3, R^4 = Me, Ph, 4-FC$_6$H$_4$, 4-ClC$_6$H$_4$,
4-MeC$_6$H$_4$; R^5 = R^6 = Me, H

88, 52-97%

Scheme 2.23 Tf$_2$NH-catalyzed reactions of VDCPs **1** with activated alkenes

a 1,4-nucleophilic addition reaction with C1′ position of VDCP **1b** to give intermediate **90a**. Subsequently, ring opening of the cyclopropyl ring in intermediate **90a** produces intermediate **91a**. Finally, products **88a** will be achieved from intermediate **91a** through intramolecular nucleophilic attack (Scheme 2.24).

In continuing research, ethyl 5,5-diarylpenta-2,3,4-trienoates **92** were also used as the reaction partner catalyzed by Brønsted acid Tf$_2$NH. It was found that cascade cycloaddition and Friedel–Crafts reactions of VDCPs **1** were achieved to provide a variety of novel polycyclic ester derivatives **93** and/or **94**, depending on the substituents R^1 and R^2 on VDCPs **1**, in moderate to good yields under mild conditions (Scheme 2.25) [41].

Plausible mechanism for these reactions is shown below: similarly, 5,5-diphenylpenta-2,3,4-trienoate **92a** is first protonated by the Brønsted acid catalyst Tf$_2$NH to produce intermediate **95a**, which will isomerize to the carbocationic intermediate **97a** via an enolate intermediate **96a**. Subsequently, VDCPs **1** undergoes a nucleophilic attack on intermediate **97a** to generate intermediate **98**, which produces intermediate **99** via the ring opening of the cyclopropyl ring. An intramolecular nucleophilic attack generates an intermediate **100**, which should

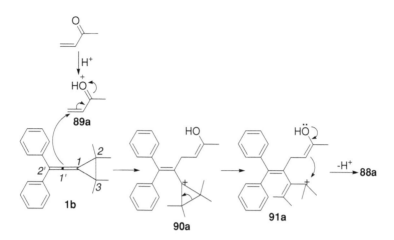

Scheme 2.24 Plausible mechanism for the Brønsted acid-catalyzed reaction of VDCP **1b** with MVK. Reprinted with the permission from Ref. [40]. Copyright 2011 American Chemical Society

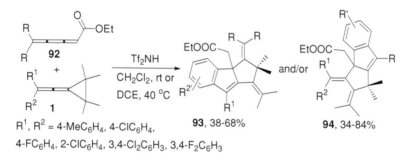

Scheme 2.25 Tf$_2$NH-catalyzed reactions of VDCPs **1** with 5,5-diarylpenta-2,3,4-trienoates **92**

be in equilibrium with intermediate **100′** via allylic rearrangement. It is understandable that the intramolecular Friedel–Crafts reaction is favored for the electron-rich rings of intermediate **100**. Thus, there are two pathways for the intramolecular Friedel–Crafts reaction. For VDCPs **1** bearing moderately electron-donating groups on the phenyl rings or neutral phenyl ring, the intramolecular Friedel–Crafts reaction occurs on the aromatic ring of VDCPs **1** to give the products **93** since R^1 and R^2 are in larger conjugate system which can stabilize the cationic intermediate generated in the Friedel–Crafts reaction (Scheme 2.26, **path a**). On the other hand, for VDCPs **1** bearing electron-withdrawing groups on both of the phenyl rings, the intramolecular Friedel–Crafts reaction occurs on the phenyl ring of 5,5-diphenylpenta-2,3,4-trienoate **92a** to generate products **94** (Scheme 2.26, **path b**).

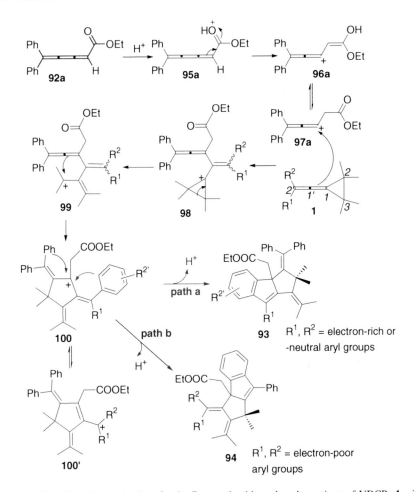

Scheme 2.26 Plausible mechanism for the Brønsted acid-catalyzed reactions of VDCPs **1** with 5,5-diphenylpenta-2,3,4-trienoate **92a**. Reprinted with the permission from Ref. [41]. Copyright 2011 Royal Society of Chemistry

2.6 Brønsted Acid-Mediated Reactions of VDCPs with Nitriles

The Brønsted acid TfOH-catalyzed reactions of VDCPs **1** with MeCN were also investigated in Shi's group and it was reported that the [3 + 2] cycloaddition products, the 3,4-dihydro-2*H*-pyrrole derivatives **101** can be obtained in moderate to excellent yields under reflux within a short time (Scheme 2.27) [1, 42–44]. In these reactions, the four substituents on the cyclopropyl ring of VDCPs **1** should be all methyl groups.

Plausible mechanism for this transformation is outlined in Scheme 2.28. First, there is an equilibrium among intermediates **102a**, **103a**, and **104a** in the reaction

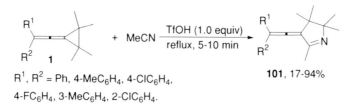

R^1, R^2 = Ph, 4-MeC$_6$H$_4$, 4-ClC$_6$H$_4$,
4-FC$_6$H$_4$, 3-MeC$_6$H$_4$, 2-ClC$_6$H$_4$.

Scheme 2.27 TfOH-catalyzed reactions of VDCPs **1** with acetonitrile. Reprinted with the permission from Ref. [1]. Copyright 2011 American Chemical Society

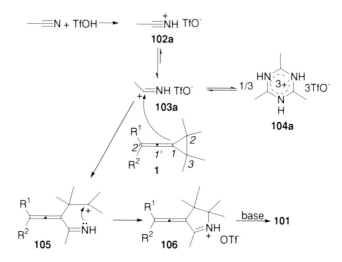

Scheme 2.28 Plausible mechanism for the TfOH-catalyzed reactions of VDCPs **1** with acetonitrile. Reprinted with the permission from Ref. [1]. Copyright 2011 American Chemical Society

of acetonitrile with TfOH according to the previous literatures [45, 46]. Intermediate **103a** undergoes an electrophilic attack to the C1 position of VDCPs **1** to afford the corresponding ring-opened cationic intermediate **105**, which undergoes the subsequent intramolecular cyclization reaction to give cationic intermediate **106**. Treatment of **106** with a base furnishes products **101**.

For strongly electron-donating 4-methoxyphenyl group substituted VDCP **1e** (R^1 = R^2 = 4-MeOC$_6$H$_4$, R^3 = R^4 = R^5 = R^6 = Me), 3,4-dihydro-2H-pyrrole derivatives **107** were formed in moderate to high yields under the same reaction conditions (Scheme 2.29).

Scheme 2.30 shows the plausible mechanism for this [3 + 2] cycloaddition reaction of VDCP **1e** with nitriles mediated by TfOH. Electrophilic attack of intermediate **103** to the C1′ position of VDCP **1e** gives intermediate **108**, which immediately undergoes a ring-opening process to afford intermediate **109**.

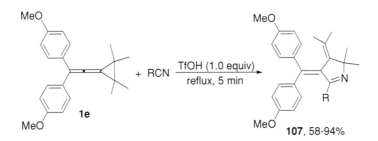

Scheme 2.29 TfOH-mediated reactions of VDCP **1e** with nitriles. Reprinted with the permission from Ref. [1]. Copyright 2011 American Chemical Society

Intramolecular cyclization takes place to give intermediate **110**, which furnishes products **107** by treatment with a base. Maybe the strong electron-donating 4-methoxyphenyl group on VDCP **1e** increases the electron density at C1′ position, facilitating the electrophilic attack of cationic intermediate **103**. Therefore, the reaction takes place in a different pathway.

2.7 Lewis Acid-Mediated Reactions of VDCPs with Acyl Chlorides

During the investigation on the Lewis acid-mediated chemistry of VDCPs **1**, it was also reported that AlCl₃-mediated tandem Friedel–Crafts reaction of VDCPs **1** with acyl chlorides **111** afforded the corresponding products **112** or **113** in moderate to good yields under mild conditions within short reaction time (Scheme 2.31) [1, 47, 48]. The control experiment showed that products **112** and **113** can be derived from the corresponding intramolecular rearrangement products of VDCPs **1**.

2.8 Lewis Acid-Mediated Reactions of VDCPs with Alcohols or Ethers

The reactions between VDCPs **1** and 1,1,3-triarylprop-2-yn-1-ols **114** or their methyl ethers **115** in the presence of Lewis acid were also investigated in Shi's group. 4-Dihydro-1*H*-cyclopenta[*b*]-naphthalene derivatives **116** can be obtained in the reactions of VDCPs **1** with 1,1,3-triarylprop-2-yn-1-ols **114** in the presence of Zr(OTf)₄ in DCE at −20 °C; 1,2,3,8-tetrahydrocyclopenta[*a*]indene derivatives **117** can be formed in the reactions of VDCPs **1**, bearing four methyl groups on the cyclopropyl ring, with 1,1,3-triarylprop-2-yn-1-ol methyl ethers **115** in the presence of Sc(OTf)₃ in DCE at 40 °C (Scheme 2.32) [49].

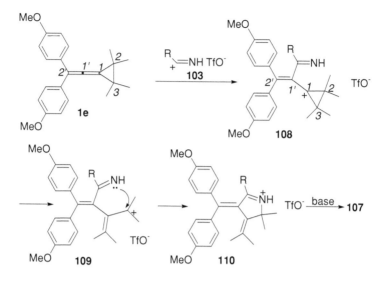

Scheme 2.30 Plausible mechanism for the TfOH-mediated reactions of VDCP **1e** with nitriles. Reprinted with the permission from Ref. [1]. Copyright 2011 American Chemical Society

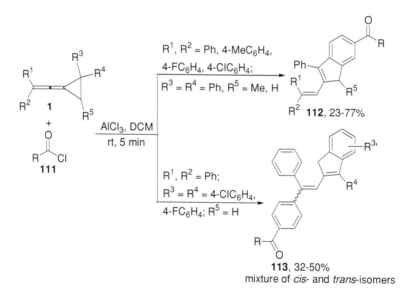

Scheme 2.31 AlCl₃-catalyzed reactions of VDCPs **1** with acyl chlorides. Reprinted with the permission from Ref. [1]. Copyright 2011 American Chemical Society

Plausible mechanism for the formation of products **116** and **117** is outlined below based on Meyer-Schuster rearrangement [50, 51]. In the presence of a Lewis acid,

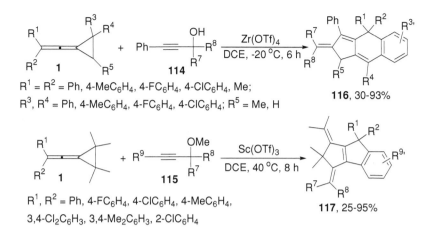

Scheme 2.32 Lewis acid-catalyzed reactions between VDCPs **1** and 1,1,3-triarylprop-2-yn-1-ols or their methyl ethers

propargylic cation intermediate **118** is first produced from 1,1,3-triarylprop-2-yn-1-ols **114** or their methyl ethers **115**. Then nucleophilic attack of C1′ position of VDCPs **1** to intermediate **118**, along with allylic rearrangement affords cationic intermediate **119**. Cyclization of intermediate **119** produces cationic intermediate **120** or its resonance-stabilized intermediate **121**. When $R^3 = R^4 = R^5 = R^6 = Me$, a Friedel–Crafts reaction with the adjacent aromatic group takes place to afford products **117**. When R^3 and R^4 are both aromatic groups, a Friedel–Crafts reaction with R^3 takes place, presumably due to steric effects, to afford products **116** (Scheme 2.33). In this case, the formation of a stable cationic intermediate **118** is the key step, so when R^7, R^8, or R^9 are aliphatic groups, complex mixtures of products are formed.

Moreover, it was found that for the reactions of VDCPs **1** with **114a**, in which both of R^7 and R^8 are 4-methoxyphenyl groups, a novel functionalized methylenecyclobutene derivatives **122** were achieved in moderate to high yields instead of the 1,2,3,8-tetrahydrocyclopenta[a]indene derivatives **117**. This may be because the intermediate formed, cation **123**, produces allyl cationic intermediate **124**, which can be further stabilized by two electron-rich aromatic groups, through an intramolecular proton transfer. Subsequent cyclization and deprotonation afford products **122** (Scheme 2.34).

Encouraged by these results, Shi et al. further explored such cascade electrophilic attack, followed by Friedel–Crafts reaction process of VDCPs **1** with other electrophiles such as enynols **126**. Using Nd(OTf)$_3$ as the catalyst, tricyclic compounds **127** can be formed in acceptable to high yields in these cases (Scheme 2.35) [52].

Plausible mechanism for the formation of tricyclic products **127a** is shown in Scheme 2.36. First, the reaction of enynol **126a** with Nd(OTf)$_3$ generates cationic intermediate **128a**, which can transform to its resonant cationic intermediate **129a**. The reaction of **129a** with VDCP **1b** gives the corresponding cyclopropyl

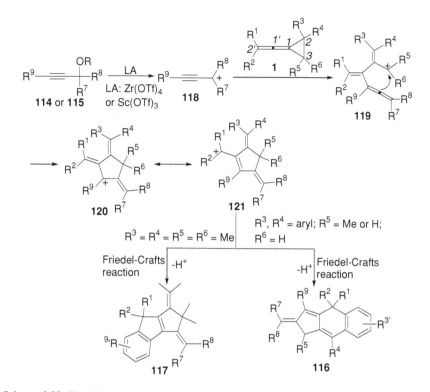

Scheme 2.33 Plausible mechanism for the Lewis acid-catalyzed reactions of VDCPs **1** with 1,1,3-triarylprop-2-yn-1-ols **114** or their methyl ethers **115**

ring-opened π-allylic cationic intermediate **130a**. In this case, the reaction of intermediate **129a** with VDCP **1b** can take place more easily than that of intermediate **128a** presumably because intermediate **129a** is less sterically hindered than intermediate **128a**. Intermediate **130a**, through intramolecular electrophilic attack, affords intermediate **131a**, which undergoes intramolecular Friedel–Crafts reaction to give the final product **127a**.

The reactions were further extended to enol **132a** and dienol **133a**, and the [3 + 2] cycloaddition products **134a** and **135a** were obtained with the reactions of VDCP **1b** (Scheme 2.37).

Plausible mechanism for the formation of products **134a** and **135a** is depicted below: first, treatment of **132a** or **133a** with Nd(OTf)$_3$ gives cationic intermediate **136**, which will transform to its resonant intermediate **137** via allylic rearrangement. Subsequently, the reaction of intermediate **137** with VDCP **1b** affords the cyclopropyl ring-opened π-allylic cationic intermediate **138**, which can either furnish intermediate **139a** (n = 1) or intermediate **140a** (n = 2) via intramolecular electrophilic attack. Nucleophilic attack by the in situ generated H$_2$O at intermediate **139a** affords the final product **134a**. Alternatively, intermediate **140a** undergoes deprotonation to afford the final product **135a** (Scheme 2.38).

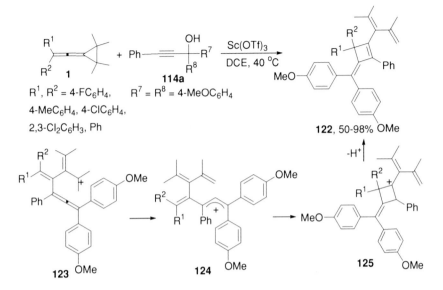

Scheme 2.34 Lewis acid-catalyzed reactions of VDCPs **1** with 1,1,3-triarylprop-2-yn-1-ol **114a**

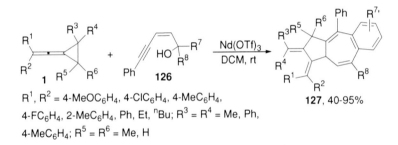

Scheme 2.35 Nb(OTf)$_3$-catalyzed reactions of VDCPs **1** with enynols **126**

When N-(4-hydroxy-4,4-diarylbut-2-ynyl)-4-methyl-N-prop-2-ynylbenzene-sulfonamides (1,6-diynes) **141** and N-allyl-N-(4-hydroxy-4,4-diarylbut-2-ynyl) -4-methylbenzenesulfonamides (1,6-enynes) **142** were tested as the partners, the reactions with VDCPs **1** can produce polycyclic compounds **143** and **144** as well as isopropylidene-3,3-diarylcyclobut-1-enyl-methyl derivatives **145** in good to high yields depending on the substituents on VDCPs **1** and substrates **141** and **142** [53]. For instance, Lewis acid Sn(OTf)$_2$ or BF$_3$·OEt$_2$-catalyzed reactions of VDCPs **1**, bearing four methyl groups on the cyclopropyl ring, with 1,6-diynes **141** and 1,6-enynes **142** can afford polycyclic compounds **143** in moderate to high yields; in particular, VDCPs **1** bearing two phenyl groups at one carbon of the cyclopropyl ring produced the corresponding polycyclic compounds **144** in

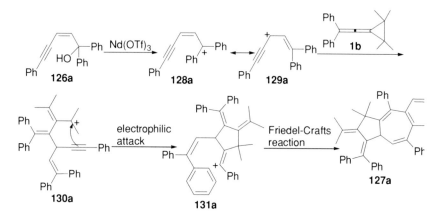

Scheme 2.36 Plausible mechanism for the Nb(OTf)$_3$-catalyzed reaction of VDCP **1b** with enynol **126a**

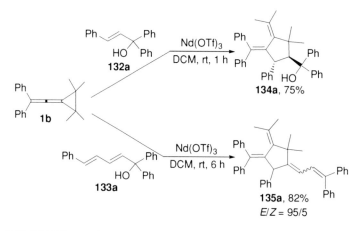

Scheme 2.37 Nb(OTf)$_3$-catalyzed reactions of VDCP **1b** with enol **132a** and dienol **133a**

51–60% yields. Moreover, the reactions of VDCPs **1** with substrates **141** or **142** bearing strong electron-donating substituents can give products **145** in moderate to high yields (Scheme 2.39).

   Plausible mechanism for the formation of products **143** is outlined below based on a cascade rearrangement [54, 55]: Lewis acid (LA)-activating **141** or **142** will give cationic intermediate **146**, which adds to the C1′ position of VDCPs **1** to afford cationic intermediate **147**. Cyclization of intermediate **147** produces

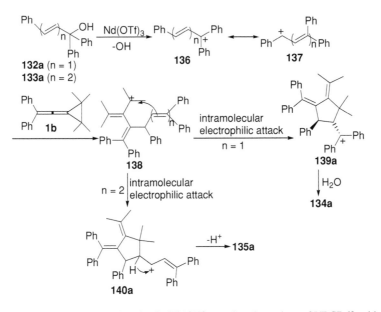

Scheme 2.38 Plausible mechanism for the Nb(OTf)$_3$-catalyzed reactions of VDCP **1b** with enol **132a** and dienol **133a**

intermediate **148**, which affords intermediate **149** via the intramolecular Friedel–Crafts reaction with the adjacent aromatic R^3 group. Aromatization of intermediate **149** furnishes polycyclic compounds **143** (Scheme 2.40).

Plausible mechanism for the formation of products **144** is indicated in Scheme 2.41. Similarly, intermediates **146a**, **150** and **151** will be formed in the Lewis acid-catalyzed reactions of **141a** with VDCPs **1**. In this case, the allylic rearrangement of intermediate **151** gives intermediate **152**, which is further stabilized by the two aryl groups and undergoes intramolecular Friedel–Crafts reaction with the adjacent phenyl ring to afford products **144**. The intramolecular Friedel–Crafts reaction with the adjacent phenyl ring of intermediate **152** can take place more easily than that of intermediate **151** since the cyclic cation in intermediate **151** is a sterically tight species which should be more difficult to go through an intramolecular Friedel–Crafts reaction. This is the reason why products **144** were obtained solely in the reactions of VDCPs **1** with two phenyl groups at one carbon of the cyclopropyl ring.

In the case of strong electron-donating groups substituted **141** (R^3 = R^4 = strong electron-donating phenyl rings) were used, similarly, intermediate **154** is formed firstly. Then intermediate **154** will transform to intermediate **155** via an intramolecular proton transfer, which can be stabilized by two electron-rich aromatic groups (R^3 and R^4). Finally, the intramolecular cyclization and deprotonation affords products **145** (Scheme 2.42).

VDCPs **1** can also react with xanthydrol in the presence of BF$_3$·OEt$_2$ to give the corresponding conjugated triene derivatives **157** in moderate to high yields under mild conditions (Scheme 2.43) [56, 57].

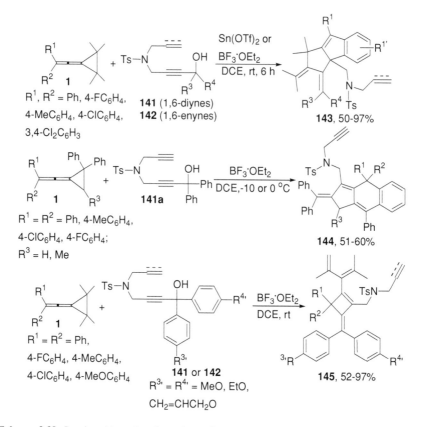

Scheme 2.39 Lewis acid-catalyzed reactions of VDCPs **1** with 1,6-diynes and 1,6-enynes

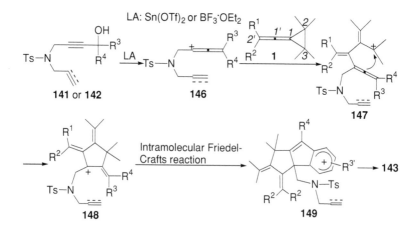

Scheme 2.40 Plausible mechanism for the formation of products **143**

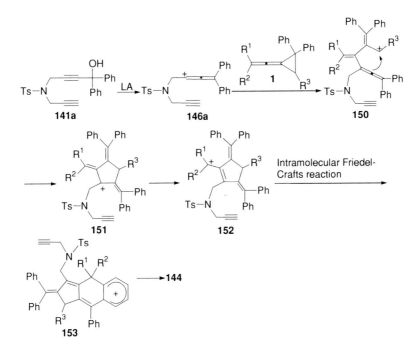

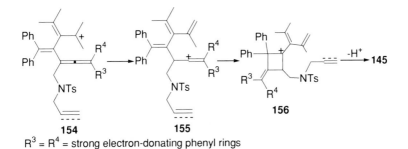

Scheme 2.41   Plausible mechanism for the formation of products **144**

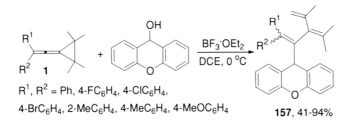

R^3 = R^4 = strong electron-donating phenyl rings

Scheme 2.42   Plausible mechanism for the formation of products **145**

R^1, R^2 = Ph, 4-FC$_6$H$_4$, 4-ClC$_6$H$_4$,
4-BrC$_6$H$_4$, 2-MeC$_6$H$_4$, 4-MeC$_6$H$_4$, 4-MeOC$_6$H$_4$ **157**, 41-94%

Scheme 2.43   BF$_3$·OEt$_2$-catalyzed reactions of VDCPs **1** with xanthydrol

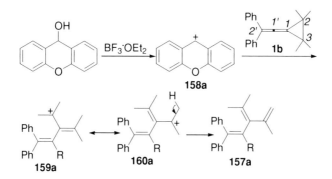

Scheme 2.44 Plausible mechanism for the BF$_3$·OEt$_2$-catalyzed reactions of VDCPs **1** with xanthydrol. Reprinted from Tetrahedron, Vol 66, Wei Yuan, Min Shi, Reactions of vinylidene-cyclopropanes with xanthydrol and xanthenes, Pages 7104–7111, Copyright 2011, with permission from Elsevier

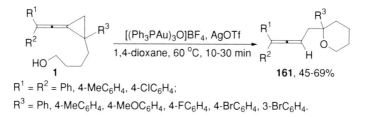

R^1 = R^2 = Ph, 4-MeC$_6$H$_4$, 4-ClC$_6$H$_4$;

R^3 = Ph, 4-MeC$_6$H$_4$, 4-MeOC$_6$H$_4$, 4-FC$_6$H$_4$, 4-BrC$_6$H$_4$, 3-BrC$_6$H$_4$.

Scheme 2.45 Lewis acid-catalyzed intramolecular ring-opening reaction of VDCPs **1** tethered with alcohol chains

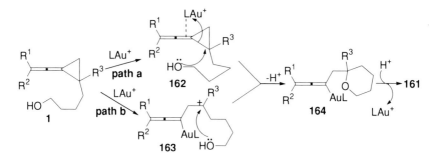

Scheme 2.46 Plausible mechanism for the formation of products **161**. Reprinted with the permission from Ref. [58]. Copyright 2011 American Chemical Society

Plausible mechanism for the reaction of VDCPs **1** with xanthydrol catalyzed by BF$_3$·OEt$_2$ is outlined in Scheme 2.44 using VDCP **1b** as the model. Initially, intermediate **158a** is formed by treatment of xanthydrol with BF$_3$·OEt$_2$, which adds to the C1′ position of VDCP **1b** to give the cyclopropyl ring-opened resonance-

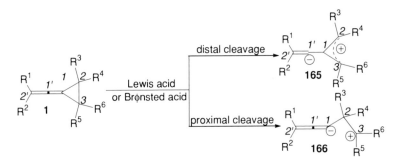

Scheme 2.47 Brief summary for the Lewis or Brønsted acid-mediated reactions of VDCPs **1**. Reprinted with the permission from Ref. [1]. Copyright 2011 American Chemical Society

stabilized intermediates **159a** and **160a** (allylic cation). Deprotonation of intermediate **160a** affords trienes **157a**.

In 2010, an intramolecular ring-opening reaction of VDCPs **1** tethered with alcohol chains was also established. With the combination of $[(Ph_3PAu)_3O]BF_4$ and AgOTf as the catalyst, allene-functionalized tetrahydropyrans **161** can be formed in moderate yields (Scheme 2.45) [58].

Plausible mechanism for the formation of products **161** is shown in Scheme 2.46. First, Au(I) complex is oxidized to L-Au$^+$ species by cocatalyst AgOTf prior to attacking VDCPs **1**. The subsequent addition/ring-opening processes can occur through two possible pathways: gold cation works as a Lewis acid to activate the allene functionality, affording intermediate **162** (**path a**) [59–62]; Au(I) catalyzes the ring opening via formation of cationic intermediate **163** (**path b**). The intramolecular nucleophilic addition by the hydroxyl group onto the electrophilic carbon centers in intermediates **162** or **163** gives the same intermediate **164**, which produces the corresponding tetrahydropyran derivatives **161** followed by protonation along with the regeneration of Au(I) species for the catalytic cycles.

In summary, the reaction course of VDCPs **1** promoted by Lewis or Brønsted acid can be categorized into the following two patterns: the distal bond and proximal bond cleavage. As can be seen from Scheme 2.47, zwitterionic ions **165** and **166** were formed with these two bond cleavage patterns. To obtain stable zwitterionic ions such as **165** and **166**, the substituents as R^3, R^4, R^5, and R^6 on the cyclopropyl ring cannot be hydrogen atoms at the same time in most cases [1].

References

1. Shi M, Shao LX, Lu JM, Wei Y, Mizuno K, Maeda H (2010) Chemistry of vinylidenecyclopropanes. Chem Rev 110:5883–5913
2. Pasto DJ, Miles MF (1976) Electrophilic addition reactions of alkenylidenecyclopropanes. Formation of highly substituted, nonplanar butadienes. J Org Chem 41:425–432

3. Pasto DJ, Gontarz JA (1969) The mechanism of the reduction of organomercurials with sodium borohydride. J Am Chem Soc 91:719–721
4. Gray GA, Jackson WR (1969) Sodium borohydride reduction of oxymercury compounds. J Am Chem Soc 91:6205–6207
5. Whitesides GM, Filippo JS Jr (1970) The mechanism of reduction of alkylmercuric halides by metal hydrides. J Am Chem Soc 92:6611–6624
6. DePuy CR, Van Lanen RJ (1974) Reactions of cyclopropanols with halogenating agents and other electrophiles. J Org Chem 39:3360–3365
7. Pasto DJ, Smorada RL, Turini BL, Wampfler DJ (1976) Electrophilic and radical addition reactions of a bisalkylidenecyclopropane. J Org Chem 41:432–438
8. Xu GC, Ma M, Liu LP, Shi M (2005) A novel rearrangement of arylvinylidenecyclopropanes to naphthalene derivatives catalyzed by Lewis acids or Brønsted acids. Synlett 1869–1872
9. Xu GC, Liu LP, Lu JM, Shi M (2005) Lewis acid-catalyzed rearrangement of multi-substituted arylvinylidenecyclopropanes. J Am Chem Soc 127:14552–14553
10. Zhang YP, Lu JM, Xu GC, Shi M (2007) Lewis-acid-catalyzed rearrangement of arylvinylidenecyclopropanes: significant influence of substituents and electronic nature of aryl groups. J Org Chem 72:509–516
11. Lu JM, Shi M (2007) Montmorillonite K-10-catalyzed intramolecular rearrangement of vinylidenecyclopropanes. Tetrahedron 63:7545–7549
12. Siehl HU, Aue DH (1997) In: Rappoport Z, Stang PJ (eds) Dicoordinated carbocations. Wiley, New York, pp 137–138
13. The stabilizing effect of cyclopropyl substituents on carbocations was well documented, see: Olah GA, Reddy VP, Surya Prakash GK (1992) Long-lived cyclopropylcarbinyl cations. Chem Rev 92:69–95
14. Bollinger JM, Brinich JM, Olah GA (1970) Stable carbonium ions. XCVI. Propadienylhalonium ions and 2-haloallyl cations. J Am Chem Soc 92:4025–4033
15. Carey FA, Sundberg RJ (1998) Advanced organic chemistry, 5th edn. Plenum Press, New York, pp 221, 419
16. Carey FA, Tremper HS (1969) Carbonium ion-silane hydride transfer reactions. III. Cyclopropylmethyl cations. J Am Chem Soc 91:2967–2972
17. Fleming I (2001) Improving the Friedel–Crafts reaction. Chemtracts: Org Chem 14:405–406
18. For the mechanism of the 1,3-proton shift, see: Carey FA, Sundburg RJ (1990) Advanced organic chemistry. 3rd edn. Plenum Press, New York, pp 609–613
19. Shi M, Wu L, Lu JM (2008) Gold(I)-catalyzed intramolecular rearrangement of vinylidenecyclopropanes. J Org Chem 73:8344–8347
20. Li W, Shi M, Li YX (2009) Brønsted acid mediated novel rearrangement of diarylvinylidenecyclopropanes and mechanistic investigations based on DFT calculations. Chem Eur J 15:8852–8860
21. Huang X, Su CL, Liu QY, Song YT (2008) A facile access to medium- and large-size naphthalenacarbocycles via Lewis acid mediated ring-expansion reaction of bicyclic vinylidenecyclopropanes. Synlett 229–232
22. Lu JM, Shi M (2006) Lewis acid catalyzed reaction of arylvinylidenecyclopropanes with acetals: a facile synthetic protocol for the preparation of indene derivatives. Org Lett 8:5317–5320
23. Su CL, Liu QY, Ni Y, Huang X (2009) An efficient synthesis of polysubstituted tetrahydrofuran and indene derivatives via the Lewis acid-mediated cycloaddition of VCPs with aldehydes. Tetrahedron Lett 50:4381–4383
24. Su CL, Huang X (2009) Lewis acid-mediated selective cycloadditions of vinylidenecyclopropanes with aromatic aldehydes: an efficient protocol for the synthesis of benzo[c]fluorene, furan and furo[2,3-b]furan derivatives. Adv Synth Catal 351:135–140
25. Su CL, Huang X, Liu QY, Huang X (2009) Facile synthesis of tetrahydrofurans from cycloadditions of vinylidenecyclopropanes with aldehydes and further transformations for the construction of furan, indene, and benzo[c]fluorene derivatives. J Org Chem 74:8272–8279

26. Lu JM, Shi M (2008) Lewis acid catalyzed reactions of vinylidenecyclopropanes with activated carbon-oxygen double bond: a facile synthetic protocol for functionalized tetrahydrofuran and 3,6-dihydropyran derivatives. J Org Chem 73:2206–2210

27. Lu JM, Shi M (2007) Lewis acid catalyzed reaction of arylvinylidenecyclopropanes with ethyl (arylimino)acetates: a facile synthetic protocol for pyrrolidine and 1,2,3,4-tetrahydroquinoline derivatives. Org Lett 9:1805–1808

28. When N-aryl substituted imines were used as the substrates, similar transformations can be obtained in very low yields, see: Stepakov AV, Larina AG, Molchanov AP, Stepakova LV, Starova GL, Kostikov RR (2007) Reaction of vinylidenecyclopropanes with aromatic imines in the presence of Lewis acids. Russ J Org Chem 43:40–49

29. Prato and Scorrano's group reported $BF_3 \cdot OEt_2$-catalyzed cycloaddition reaction of aryliminoacetates with electron-rich olefins to give tetrahydroquinoline derivatives, see: Borrione E, Prato M, Scorrano G, Stivanello M, Lucchini V (1988) Synthesis and cycloaddition reactions of ethyl glyoxylate imines. Synthesis of substituted furo-[3,2-c]quinolines and 7H-indeno[2,1-c]quinolines. J Heterocycl Chem 25:1831–1835

30. Regás D, Afonso MM, Rodríguez ML, Antonio Palenzuela J (2003) Synthesis of octahydroquinolines through the Lewis acid catalyzed reaction of vinyl allenes and imines. J Org Chem 68:7845–7852

31. Hayashi Y, Shibata T, Narasaka K (1990) Ene reaction of allenyl sulfides with aldehydes and Schiffs bases catalyzed by Lewis acids. Chem Lett 1693–1696

32. Xu B, Shi M (2003) Lewis acid-catalyzed reaction of allenes with activated ketone. Synlett 1639–1642

33. Chevrier B, Weiss R (1974) Structures of the intermediate complexes in Friedel–Crafts acylations. Angew Chem Int Ed Engl 13:1–10

34. Kobayashi has concluded that this type of aza-Diels–Alder reaction proceeded via a stepwise mechanism, see: Kobayashi S, Ishitani H, Nagayama S (1995) Lanthanide triflate catalyzed imino Diels–Alder reactions; convenient syntheses of pyridine and quinoline derivatives. Synthesis 1195–1202

35. Shi M, Shao LX, Xu B (2003) The Lewis acids catalyzed aza-Diels–Alder reaction of methylenecyclopropanes with imines. Org Lett 5:579–582

36. Shao LX, Shi M (2003) Montmorillonite KSF-catalyzed one-pot, three-component, aza-Diels–Alder reactions of methylenecyclopropanes with arenecarbaldehydes and arylamines. Adv Synth Catal 345:963–966

37. Zhu ZB, Shao LX, Shi M (2009) Brønsted acid or solid acid catalyzed aza-Diels–Alder reactions of methylenecyclopropanes with ethyl (arylimino)acetates. Eur J Org Chem 2576–2580

38. Lu JM, Zhu ZB, Shi M (2009) Lewis acid or Brønsted acid catalyzed reactions of vinylidene cyclopropanes with activated carbon-nitrogen, nitrogen–nitrogen, and iodine-nitrogen double-bond-containing compounds. Chem Eur J 15:963–971

39. Li Z, Conser KR, Jacobsen EN (1993) Asymmetric alkene aziridination with readily available chiral diimine-based catalysts. J Am Chem Soc 115:5326–5327

40. Li W, Shi M (2009) A catalytic method for the preparation of polysubstituted cyclopentanes: [3 + 2] cycloaddition of vinylidenecyclopropanes with activated olefins catalyzed by triflic imide. J Org Chem 74:856–860

41. Li W, Shi M (2009) Triflic imide-catalyzed cascade cycloaddition and Friedel–Crafts reaction of diarylvinylidenecyclopropanes with ethyl 5,5-diarylpenta-2,3,4-trienoate. Org Biomol Chem 7:1775–1777

42. Li W, Shi M (2008) Brønsted acid TfOH-mediated [3 + 2] cycloaddition reactions of diarylvinylidenecyclopropanes with nitriles. J Org Chem 73:4151–4154

43. For related results of methylenecyclopropanes from Shi's group, see: Huang JW, Shi M (2004) Brønsted acid TfOH-mediated reactions of methylenecyclopropanes with nitriles. Synlett 2343–2346 and references therein

44. Shi M, Tian GQ (2006) Brønsted acid TfOH-mediated reactions of 2-(arylmethylene) cyclopropylcarbinols with acetonitrile. Tetrahedron Lett 47:8059–8062

45. Booth BL, Noori GFM (1980) The chemistry of nitrilium salts. Part 1. Acylation of phenols and phenol ethers with nitriles and trifluoromethanesulphonic acid. J Chem Soc Perkin Trans 1:2894–2900
46. Amer MI, Booth BL, Noori GFM, Proença MFJRP (1983) The chemistry of nitrilium salts. Part 3. The importance of triazinium salts in Houben-Hoesch reactions catalyzed by trifluoromethanesulphonic acid. J Chem Soc Perkin Trans 1:1075–1082
47. Shi M, Wu L, Lu JM (2008) AlCl₃-mediated tandem Friedel–Crafts reactions of vinylidenecyclopropanes with acyl chlorides: a facile synthetic method for the construction of 1-[2-(2,2-diarylvinyl)-1-phenyl-3H-inden-5-yl]ethanone derivatives. Tetrahedron 64: 3315–3321
48. For pioneering work on the reactions of methylenecyclopropanes with acyl chloride in the presence of AlCl₃, see: Huang X, Yang YW (2007) Acylation of alkylidenecyclpropanes for the facile synthesis of α,β-unsaturated ketone and benzofulvene derivatives with high stereoselectivity. Org Lett 9:1667–1670
49. Shi M, Yao LF (2008) Lewis acid catalyzed reactions of diarylvinylidenecyclopropanes and 1,1,3-triarylprop-2-yn-1-ols or their methyl ethers. Chem Eur J 14:8725–8731
50. Swaminathan S, Narayanan KV (1971) The Rupe and Meyer-Schuster rearrangements. Chem Rev 71:429–438
51. Edens M, Boerner D, Chase CR, Nass D, Schiavelli MD (1977) Mechanism of the Meyer-Schuster rearrangement. J Org Chem 42:3403–3408
52. Yao LF, Shi M (2009) Nd(OTf)₃-catalyzed cascade reactions of vinylidenecyclopropanes with enynol: a new method for the construction of the 5-7-6 tricyclic framework and its scope and limitations. Eur J Org Chem 4036–4040
53. Yao LF, Shi M (2009) Lewis acid catalyzed cascade reactions of 1,6-diynes and 1,6-enynes with vinylidenecyclopropanes. Chem Eur J 15:3875–3881
54. Michelet V, Toullec PY, Genêt JP (2008) Cycloisomerization of 1, n-enynes: challenging metal-catalyzed rearrangements and mechanistic insights. Angew Chem Int Ed 47: 4268–4315
55. Bruneau C (2005) Electrophilic activation and cycloisomerization of enynes: a new route to functional cyclopropanes. Angew Chem Int Ed 44:2328–2334
56. Yuan W, Shi M (2010) Reactions of vinylidenecyclopropanes with xanthydrol and xanthenes. Tetrahedron 66:7104–7111
57. For pioneering work with bis(4-alkoxyphenyl)methanols as the electrophiles, please see: Wu L, Shi M, Li YX (2010) BF₃·OEt₂-catalyzed intermolecular reactions of vinylidenecyclopropanes with bis(p-alkoxyphenyl)methanols: a novel cationic 1,4-aryl-migration process. Chem Eur J 16:5163–5172
58. Li W, Yuan W, Pindi S, Shi M, Li GG (2010) Au/Ag-catalyzed intramolecular ring-opening of vinylidene-cyclopropanes (VDCPs): an easy access to functional tetrahydropyrans. Org Lett 12:920–923
59. Hyland CJT, Hegedus LS (2006) Gold-catalyzed and N-iodosuccinimide-mediated cyclization of γ-substituted allenamides. J Org Chem 71:8658–8660
60. Morita N, Krause N (2006) The first gold-catalyzed C-S bond formation: cycloisomerization of α-thioallenes to 2,5-dihydrothiophenes. Angew Chem Int Ed 45:1897–1899
61. Sromek AW, Rubina M, Gevorgyan V (2005) 1,2-Halogen migration in haloallenyl ketones: regiodivergent synthesis of halofurans. J Am Chem Soc 127:10500–10501
62. Nishina N, Yamamoto Y (2006) Gold-catalyzed intermolecular hydroamination of allenes with arylamines and resulting high chirality transfer. Angew Chem Int Ed 45:3314–3317

Chapter 3
Transition Metal-Catalyzed Transformations of VDCPs

Abstract Transition metals such as palladium and rhodium-catalyzed transformations of vinylidenecyclopropanes are introduced in this chapter.

Keywords Vinylidenecyclopropanes · Transition metal · Palladium · Rhodium · Coupling reaction

In 2006, Shi et al. reported that the Pd(0)-catalyzed reactions of VDCPs **1** with acetic acid can proceed efficiently to give the corresponding cyclopropyl ring-opened acetylated dienes **167** in moderate to good yields in the presence of bis[(2-diphenylphosphino)phenyl] ether (DPEphos) ligand under mild reaction conditions (Scheme 3.1) [1–4].

In 2009, Shi and co-workers reported the palladium-catalyzed coupling reactions of VDCPs **1** with 2-iodophenols **168** and N-(2-iodophenyl)-4-methylbenzenesulfonamide **170**. For instance, using $PdCl_2$ as the catalyst, and 1,3-bis(diphenylphosphanyl)propane (dppp) as the ligand in the presence of $^{i}Pr_2$NEt, 2,2-diaryl-3-(tetramethylcyclopropylidene)-2,3-(dihydro) benzofuran **169** can be obtained in 36–94% yields. In addition, using $PdCl_2$ as the catalyst, and dppp as the ligand in the presence of $^{i}Pr_2$NEt, Ag_2CO_3 and $[Et_3NH][BF_4]$, 2,2-diaryl-3-tetramethylcyclopropylidene-1-(toluene-4-sulfonyl)-2,3-dihydro-1H-indole derivatives **171** can be formed in 30–64% yields (Scheme 3.2) [5].

Treatment of VDCPs **1** with $PdCl_2$ can afford novel dimeric allylpalladium(II) complexes **172** or **173** depending on the substituents attached on the cyclopropyl ring of VDCPs **1**. For example, with VDCPs **1** in which both of R^3 and R^4 are methyl groups, dimeric allylpalladium(II) complexes **172** can be obtained in acceptable to moderate yields; whereas with VDCPs **1** in which R^1, R^2, R^3, and R^4 are aryl groups, dimeric allylpalldium(II) complexes **173** were formed in moderate to good yields as the major products, in which case when R^3 and R^4 as well as R^1 and R^2 are aryl groups bearing electron-donating groups, the $PdCl_2$-catalyzed

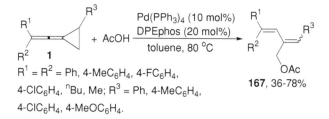

R^1 = R^2 = Ph, 4-MeC$_6$H$_4$, 4-FC$_6$H$_4$,
4-ClC$_6$H$_4$, ^{n}Bu, Me; R^3 = Ph, 4-MeC$_6$H$_4$,
4-ClC$_6$H$_4$, 4-MeOC$_6$H$_4$.

Scheme 3.1 Palladium-catalyzed reactions of VDCPs **1** with acetic acid. Reprinted with the permission from [1]. Copyright 2011 American Chemical Society

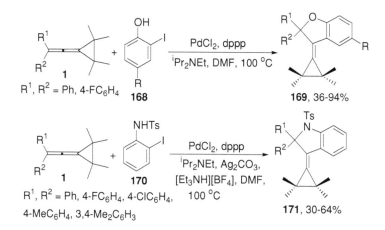

Scheme 3.2 Palladium-catalyzed coupling reactions of VDCPs **1** with iodides

rearrangement products **21** will be obtained as the minor products (Scheme 3.3) [6].

On the basis of the above results, plausible mechanism for the formation of dimeric allylpalladium(II) complexes **172** and **173** is outlined in Scheme 3.4. The coordination of PdCl$_2$ with VDCPs **1** produces the initial zwitterionic intermediate **174**, from which the corresponding cyclopropyl ring-opened zwitterionic intermediate **175** is formed. Intermediate **175** can exist as intermediate **176** and its resonance-stabilized intermediate **176′**. As for VDCPs **1** in which both of R^3 and R^4 are methyl groups and R^5 is an aryl group, intermediate **176** is a more reactive species since it bears two alkyl groups R^3 and R^4 at the C2 position, which can be easily transformed to the corresponding intermediate **177** along with the elimination of HCl. Then, the corresponding dimeric allylpalladium(II) complexes **172** are produced from intermediate **177**. On the other hand, for VDCPs **1** in which both of R^3 and R^4 are aryl groups and R^5 is methyl group, intermediate **176′** is more reactive species, which

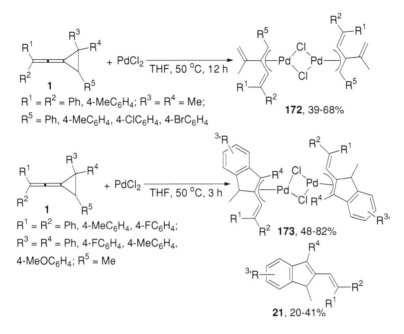

Scheme 3.3 Reactions of VDCPs **1** with PdCl$_2$

easily undergoes intramolecular Friedel–Crafts reaction with the adjacent aromatic group at the C2 position (R^3 group) to afford the corresponding intermediate **178** along with the elimination of HCl. Intermediate **178** can be transformed to the corresponding dimeric allylpalladium(II) complexes **173** as well as products **21** by protonation. If R^3 and/or R^4 are/is electron-rich aromatic group(s), the intramolecular Friedel–Crafts reaction is facilitated, so the formation of products **21** is favored in these cases.

An efficient catalytic system for the intramolecular ene reaction of alkene moiety attached VDCPs **1** was also established by Shi's group. For example, using [RhCl(CO)$_2$]$_2$ as the catalyst in co-solvent of toluene and acetonitrile, the reaction was achieved to afford bicycle[5.1.0]octylene derivatives **179** in moderate to high yields (Scheme 3.5) [7].

As in a similar mechanism proposed by Taw et al. and Prater et al. [8, 9], plausible mechanism for the formation of products **179** is outlined in Scheme 3.6. First, the terminal alkene in VDCPs **1** can be coordinated to Rh(I) metal center to generate intermediate **180**. Subsequently, the oxidative addition of Rh(I) center into the neighboring allylic C–H bond gives a π-allyl Rh-H species **181** [10–12], which leads to a seven-membered carbocyclic Rh-H species **182** through an intramolecular cycloaddition to the allene moiety. Finally, reductive elimination affords the products **179** along with the regeneration of the Rh(I) catalyst.

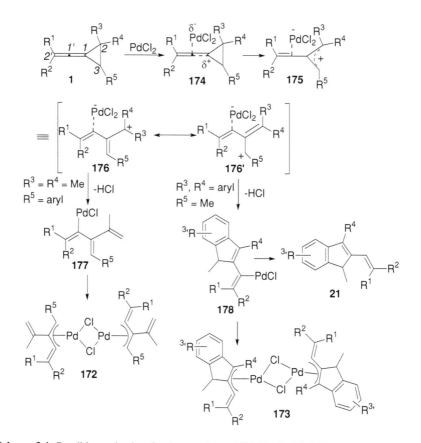

Scheme 3.4 Pausible mechanism for the reactions of VDCPs **1** with PdCl$_2$. Reprinted with the permission from [6]. Copyright 2011 American Chemical Society

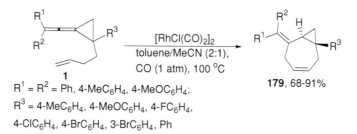

Scheme 3.5 Rhodium-catalyzed intramolecular ene reactions of VDCPs **1**

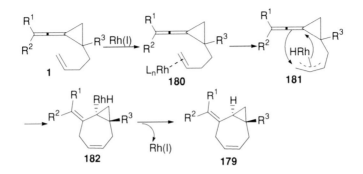

Scheme 3.6 Plausible mechanism for the rhodium-catalyzed intramolecular ene reactions of VDCPs **1**. Reprinted with the permission from [7]. Copyright 2011 American Chemical Society

References

1. Shi M, Shao LX, Lu JM, Wei Y, Mizuno K, Maeda H (2010) Chemistry of vinylidenecyclopropanes. Chem Rev 110:5883–5913
2. Lu JM, Shi M (2006) Palladium-catalyzed reactions of vinylidenecyclopropanes with acetic acid. Tetrahedron 62:9115–9122
3. Shi M, Wang BY, Huang JW (2005) Palladium-catalyzed isomerization of methylenecyclopropanes in acetic acid. J Org Chem 70:5606–5610
4. Shi M, Wang BY, Shao LX (2007) Regioselective control in the palladium-catalyzed isomerization of methylenecyclopropylcarbinols using acetic acid as a reagent. Synlett 909–912
5. Li W, Shi M (2009) Palladium-catalyzed coupling reactions of diarylvinylidenecyclopropanes with 2-iodophenol and N-(2-iodophenyl)-4-methylbenzenesulfonamide. Eur J Org Chem 270–274
6. Gu XX, Wang FJ, Qi MH, Shao LX, Shi M (2009) Synthesis of novel allylpalladium(II) complexes from PdCl$_2$-promoted ring-opening reactions of vinylidenecyclopropanes. Organometallics 28:1569–1574
7. Li W, Yuan W, Shi M, Hernandez E, Li GG (2010) Rhodium(I)-catalyzed intramolecular ene reaction of vinylidenecyclopropanes and alkenes for the formation of bicycle[5.1.0]octylenes. Org Lett 12:64–67
8. Taw FL, White PS, Bergman RG, Brookhart M (2002) Carbon-carbon bond activation of R–CN (R = Me, Ar, ^{i}Pr, ^{t}Bu) using a cationic Rh(III) complex. J Am Chem Soc 124:4192–4193
9. Prater ME, Pence LE, Clérac R, Finniss GM, Campana C, Auban-Senzier P, Jérome D, Canadell E, Dunbar KR (1999) A remarkable family of rhodium acetonitrile compounds spanning three oxidation states and with nuclearities ranging from mononuclear and dinuclear to one-dimensional chains. J Am Chem Soc 121:8005–8016
10. Wegner HA, de Meijere A, Wender PA (2005) Transition metal-catalyzed intermolecular [5 + 2] and [5 + 2+1] cycloadditions of allenes and vinylcyclopropanes. J Am Chem Soc 127:6530–6531
11. Brummond KM, Chen HF, Mitasev B, Casarez AD (2004) Rhodium(I)-catalyzed ene-allene carbocyclization strategy for the formation of azepines and oxepines. Org Lett 6:2161–2163
12. Yu ZX, Cheong PHY, Liu P, Legault CY, Wender PA, Houk KN (2008) Origins of differences in reactivities of alkenes, alkynes, and allenes in [Rh(CO)$_2$Cl]$_2$-catalyzed (5 + 2) cycloaddition reactions with vinylcyclopropanes. J Am Chem Soc 130:2378–2379

Chapter 4
Reactions of VDCPs with Electrophiles

Abstract Reactions of vinylidenecyclopropanes with a variety of electrophiles such as diaryl diselenide, iodine, bromine, N-fluorodibenzenesulfonimide (NFSI), and N-bromosuccinimide are shown in this chapter.

Keywords Vinylidenecyclopropanes · Electrophilic addition · Ring-opening · Rearrangement

An interesting addition reaction of VDCPs **1** with diaryl diselenide **183** catalyzed by iodosobenzene diacetate was first reported by Shi's group. The corresponding addition products **184** could be obtained in moderate to good yields under mild conditions (Scheme 4.1) [1, 2].

Shi et al. also investigated the reactions between VDCPs **1** and N-fluorodibenzenesulfonimide (NFSI) and the corresponding fluorinated derivatives **185** can be achieved in good to high yields (Scheme 4.2) [3].

Plausible mechanism for the reactions of VDCPs **1** with NFSI is outlined in Scheme 4.3. Initially, the fluoric cation F^+ and the anionic intermediate $(PhSO_2)_2N^-$ are generated from NFSI under the stardard reaction conditions [4–6]. Then the fluoric cation F^+ adds to the C1' position of VDCPs **1** to give cationic intermediate **186**. Intermediate **186** can give the cyclopropyl ring-opened cationic intermediate **187**, which can be transformed to the final products **185** via deprotonation.

It was also found that VDCPs **1** can undergo ring-opening reactions upon treatment with iodine or bromine at 0–25 °C in DCE to give the corresponding iodinated or brominated naphthalene derivatives **188** or **189** in low to high yields within 3 h (Scheme 4.4) [1, 7].

In the following research, it was observed that the reactions of VDCPs **1** with equimolar amount of bromine or iodine at low temperature can produce the corresponding addition products **190–192**, depending on the nature of VDCPs, in moderate to good yields at −40 and −100 °C, respectively. In addition, the

L. Shao et al., *Chemical Transformations of Vinylidenecyclopropanes*,
SpringerBriefs in Molecular Science, DOI: 10.1007/978-3-642-27573-9_4,
© The Author(s) 2012

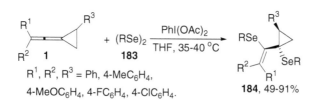

Scheme 4.1 PhI(OAc)$_2$−mediated reactions of VDCPs **1** with diaryl diselenide. Reprinted with the permission from Ref. [1]. Copyright 2011 American Chemical Society

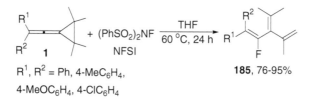

Scheme 4.2 Reactions of VDCPs **1** with NFSI

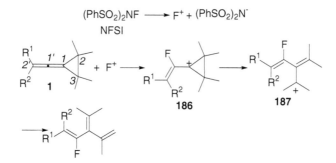

Scheme 4.3 Plausible mechanism for the reactions of VDCPs **1** with NFSI. Reprinted from Ref. [3]. Copyright 2011, with permission from Elsevier

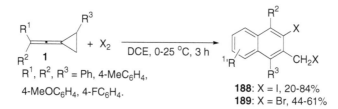

Scheme 4.4 Reactions of VDCPs **1** with I$_2$ and Br$_2$ at 0–25 °C. Reprinted with the permission from Ref. [1]. Copyright 2011 American Chemical Society

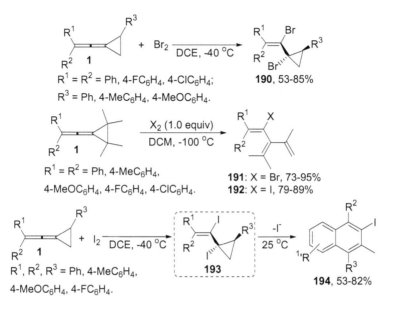

Scheme 4.5 Reactions of VDCPs **1** with I$_2$ and Br$_2$ at −40 or −100 °C. Reprinted with the permission from Ref. [1]. Copyright 2011 American Chemical Society

reactions of VDCPs **1** with equimolar amount of iodine gave the corresponding iodinated naphthalene derivatives **194** presumably derived from the corresponding addition products **193** at 25 °C (Scheme 4.5) [1, 8].

Interestingly, a drastic solvent effect was found to result in different products during the investigation on the reactions of VDCPs **1** with bromine. For example, the brominated indene derivatives **195** were obtained in good to high yields in DCM at −100 °C; however, the brominated conjugate triene derivatives **196** were obtained in diethyl ether at the same temperature (Scheme 4.6) [1, 9].

VDCPs **1** can also undergo hydrobromination or alkoxybromination in the presence of *N*-bromosuccinimide (NBS) and water or alcohols to give the corresponding vinylbromohydrin **197** and vinylbromoalkoxy derivatives **198** in moderate to excellent yields at room temperature (Scheme 4.7) [1, 10].

Huang et al. investigated the halohydroxylation reactions of VDCPs **1** carefully and they found that highly regioselective halohydroxylations of bicyclic VDCPs **1** can be achieved to give four types of products: **199–202**, depending on the reaction conditions and the size of the ring of VDCPs **1** [11]. For example, the halohydroxylation reactions of bicyclic VDCPs **1**, in the presence of 1.2 ~ 1.5 equiv. of *N*-bromosuccinimide (NBS) or *N*-iodosuccinimide (NIS), occur smoothly at room temperature to give the cyclopropyl ring-kept products **199** in moderate to high yields with high regio- and diastereo-selectivity; the reactions between bicyclic VDCPs **1** and 2.0 equiv. of NBS at 100 °C afford products **200** in 48–75% yields by proximal cleavage of the cyclopropyl ring (Scheme 4.8).

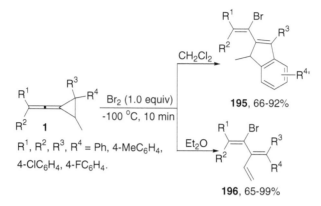

Scheme 4.6 Reactions of VDCPs **1** with Br_2 at -100 °C in different solvents. Reprinted with the permission from Ref. [1]. Copyright 2011 American Chemical Society

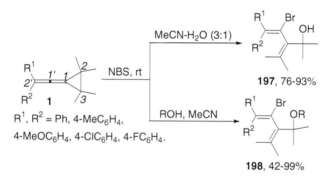

Scheme 4.7 Reactions of VDCPs **1** with NBS. Reprinted with the permission from Ref. [1]. Copyright 2011 American Chemical Society

Further studies showed that treatment of product **199a** with NBS (1.0 equiv.) at 100 °C, the ring-expansion product **200a** can be achieved in 82% yield; treatment of product **199b** with I⁺ in DMSO at room temperature for 18 h, intermediate **203a** can be obtained in 68% yield; in addition, further treatment of intermediate **203a** at 85–90 °C in the presence of Et_3N smoothly furnished **201a** in 70% yield (Scheme 4.9).

On the basis of the above control experiments and the obtained intermediate **203a**, plausible mechanism for the formation of products **199–201** is shown in Scheme 4.10 with VDCP **1f** as the model. Initially, X⁺ adds to C1′–C1 double bond of VDCP **1f** to give the halonium intermediate **204**, which is selectively attacked by H_2O from the less sterically hindered side to produce the ring-kept halohydroxylation products **199** with high diastereoselectivity. In the presence of another equivalent of NBS at 100 °C, the addition of bromonium ion to the double bond induces ring enlargement to give the dibromocyclobutanone intermediate

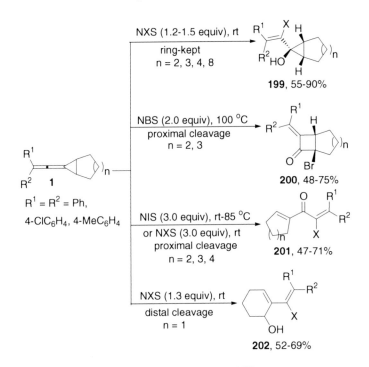

Scheme 4.8 Reactions of bicyclic VDCPs **1** with NBS or NIS

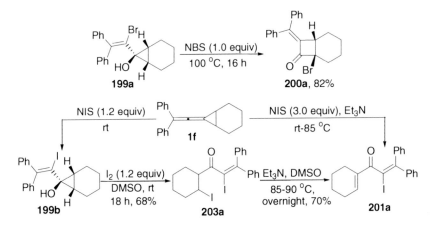

Scheme 4.9 Control experiments for the reactions of bicyclic VDCP **1f** with NBS or NIS. Reprinted with the permission from Ref. [11]. Copyright 2011 Wiley John and Sons

206a. Then a rearrangement occurs with bromine atom migration mediated by HBr to afford intermediate **207a** [12, 13], which eliminates one molecular of HBr to furnish product **200a** (**path a**). However, in the presence of I⁺, a different

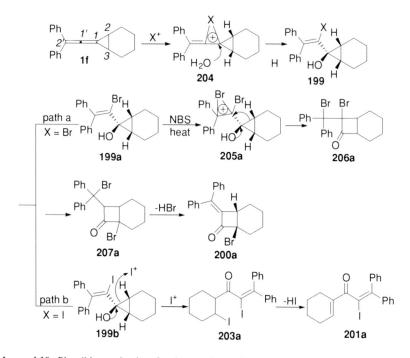

Scheme 4.10 Plausible mechanism for the reactions of VDCPs **1** with NBS or NIS. Reprinted with the permission from Ref. [11]. Copyright 2011 Wiley John and Sons

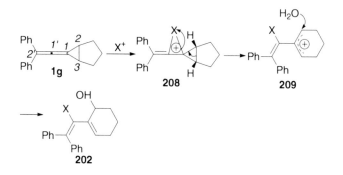

Scheme 4.11 Plausible mechanism for the formation of product **202**. Reprinted with the permission from Ref. [11]. Copyright 2011 Wiley John and Sons

electrophilic ring-opening reaction of **199b** occurs with high stereoselectivity to give intermediate **203a**, which eliminates one molecular of HI to achieve product **201a** (**path b**).

Formation of products **202** is proposed in Scheme 4.11 with VDCP **1g** as the model. Initially, X^+ adds to C1′–C1 double bond of VDCP **1g** to give the halonium ion intermediate **208**, which undergoes a ring-expansion process to afford intermediate **209**. Then attack of H_2O to intermediate **209** furnishes products **202**. Maybe the formation of intermediate **209** as a six-membered ring cation is thermally favored in this case, thus results in the different reaction pattern.

References

1. Shi M, Shao LX, Lu JM, Wei Y, Mizuno K, Maeda H (2010) Chemistry of vinylidenecyclopropanes. Chem Rev 110:5883–5913
2. Shi M, Lu JM (2005) Reactions of vinylidenecyclopropanes with diaryl diselenide catalyzed by iodosobenzene diacetate and further transformation of the adducts. Synlett 2352–2356
3. Jiang M, Shi M (2009) Reactions of methylenecyclopropanes and vinylidenecyclopropanes with N-fluorodibenzenesulfonimide. Tetrahedron 65:5222–5227
4. Eberson L, Persson O (1997) Fluoro spin adducts and their modes of formation. J Chem Soc Perkin Trans 2:893–898
5. Jung ME, Toyota A (2001) Preparation of 4′-substituted thymidines by substitution of the thymidine 5′-esters. J Org Chem 66:2624–2635
6. Hill B, Liu Y, Taylor SD (2004) Synthesis of α-fluorosulfonamides by electrophilic fluorination. Org Lett 6:4285–4288
7. Shi M, Ma M, Shao LX (2005) Ring-opening reactions of diarylvinylidenecyclopropanes by iodine and bromine. Tetrahedron Lett 46:7609–7613
8. Shi M, Ma M, Zhu ZB, Li W (2006) Reactions of arylvinylidenecyclopropanes with bromine and iodine. Synlett 1943–1947
9. Shi M, Li W (2007) Reactions of diarylvinylidenecyclopropanes with bromine at −100 °C in dichloromethane and ether. A drastic solvent effect. Tetrahedron 63:6654–6660
10. Li W, Shi M (2008) An efficient method for the synthesis of vinylbromohydrin and vinylbromoalkoxy derivatives and cyclocarbonylation of α-allenic alcohols catalyzed by palladium chloride. J Org Chem 73:6698–6705
11. Su CL, Cao J, Huang X, Wu LL, Huang X (2011) Regiocontrolled halohydroxylations of bicyclic vinylidenecyclopropanes: a versatile strategy for the construction of diverse highly functionalized carbocyclic scaffolds. Chem Eur J 17:1579–1585
12. Djerassi C, Scholz CR (1947) The bromination of 3-ketosteroids in acetic acid and the effect of trace substances in the solvent. J Am Chem Soc 69:2404–2410
13. Cabaj JE, Kairys D, Benson TR (2007) Development of a commercial process to produce oxandrolone. Org Process Res Dev 11:378–388

Chapter 5
Miscellaneous Analogs

Abstract The reactions of vinylidenecyclopropanes (VDCPs) **1**, which cannot be classified into the foregoing chapters, are shown in this chapter. In this chapter, Brønsted base-mediated as well as radical reaction, cycloaddition reaction and oxidation reaction of vinylidenecyclopropanes are displayed.

Keywords Vinylidenecyclopropanes · Brønsted base · Radical reaction · Cycloaddition reaction · Oxidation reaction

5.1 Brønsted Base-Mediated Transformations of VDCPs

VDCPs **1** can be isomerized to vinylcyclopropenes **210** in good to high yields within 5 h under basic conditions, which can also undergo Lewis acid-catalyzed rearrangement reactions to give the corresponding naphthalenes **211** or indenes **212**, respectively (Scheme 5.1) [1, 2].

Highly selective addition reactions of VDCPs **1** were realized by treatment with lithium diisopropylamide (LDA) in THF and quenching with aldehydes, ketones, and enones, respectively. A number of vinylcyclopropene derivatives **213**, triene derivatives **214**, allenol derivatives **215**, and 1,3-enyne derivatives **216** can be obtained selectively in moderate to good yields depending on the nature of the R^4 group on the cyclopropyl ring and different electrophiles (Scheme 5.2) [3, 4].

Plausible mechanism for the formation of products **213–216** is outlined in Scheme 5.3. Initially, the lithiation of cyclopropyl ring of VDCPs **1** gives the corresponding cyclopropyl carbanion intermediate **217** by treatment with LDA [5]. Intermediate **217** can be transformed to intermediates **218**, **219**, and **220** [6–11]. When both of R^3 and R^4 are aryl groups and aldehydes are used as the electrophiles, vinylcyclopropene derivatives **213** are formed through intermediate **221**

L. Shao et al., *Chemical Transformations of Vinylidenecyclopropanes,*
SpringerBriefs in Molecular Science, DOI: 10.1007/978-3-642-27573-9_5,
© The Author(s) 2012

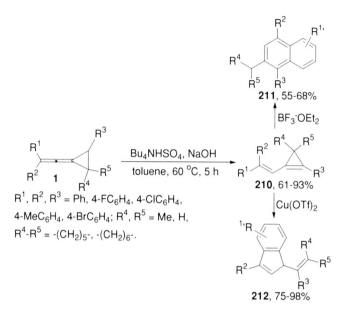

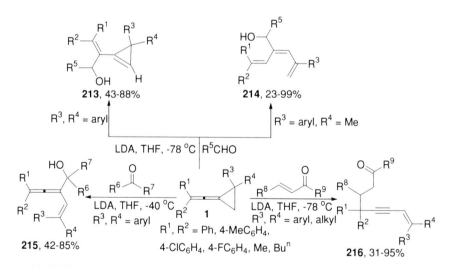

Scheme 5.2 LDA-mediated reactions of VDCPs **1** with aldehydes, ketones, and enones

(**path a**). When ketones are used as the electrophiles, allenol derivatives **215** are obtained by the reaction of intermediate **220** with ketones through intermediate **225** and it is conceivable that a six-membered transition state TS1 is concerned in

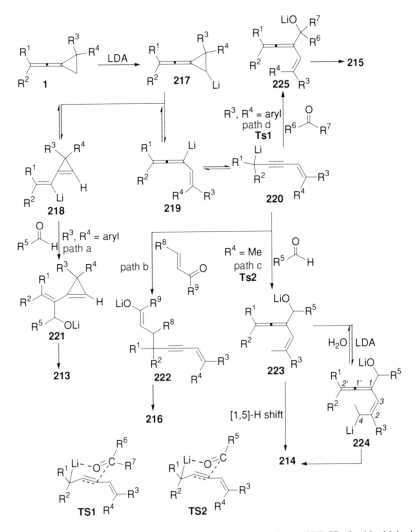

Scheme 5.3 Plausible mechanism for the LDA-mediated reactions of VDCPs **1** with aldehydes, ketones and enones

this reaction (**path d**). The different reactivity between aldehydes and ketones may be due to the steric effect of the two electrophiles. When R^4 is methyl and aldehydes are used as the electrophiles, intermediate **223** is formed by the reaction of intermediate **220** with aldehydes through a six-membered transition state TS2, which undergoes a [1,5]-H shift to produce the triene derivatives **214** [12]. When CO_2 was used as the electrophile, similar transformations of VDCPs **1** were achieved. See [13]. Alternatively, intermediate **224** can be formed by lithiation of intermediate **223** (R^4 is methyl), which undergoes a [1,5]-lithium shift to give products **214**. During the two possible pathways, the ambient water can take part in

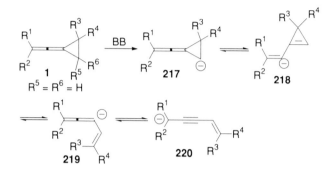

Scheme 5.4 Brief summary for the Brønsted base-mediated transformations of VDCPs **1**. Reprinted with the permission from Ref. [1]. Copyright 2011 American Chemical Society

the process to partially replace the lithium with proton, which was further confirmed by the deuterium-labeling experiment (**path c**). When enones are used as the electrophiles, intermediate **222** is formed by reaction of intermediate **220** with enones and subsequently 1,3-enyne derivatives **216** are obtained, presumably as the lithiated sp^3 carbanion prefers such Michael addition (**path b**) .When ynones were used as the electrophiles, novel domino carbolithiation reactions of VDCPs **1** were achieved. See [14].

In these reactions, there may be an equilibrium between the lithiated anions such as **217–220**. To carry out these reactions smoothly, both of the substituents on one carbon of the cyclopropyl ring of VDCPs **1** should be hydrogen atoms ($R^5 = R^6 = H$); at the same time, R^3 and R^4 can be both aryl or alkyl groups or one of them is an aryl group and the other is an alkyl group, and neither of them can be hydrogen atom (Scheme 5.4).

5.2 Radical Reactions of VDCPs

It was found that the reactions of VDCPs **1** with diphenyl diselenide **183a** could also take place in the presence of AIBN to produce the corresponding products **184** or **226** in moderate to good yields under mild conditions (Scheme 5.5) [1, 15].

VDCPs **1** can also undergo the reaction with diaryl diselenide **183** upon heating at 150 °C to give the corresponding 1,2-diarylselenocyclopentene derivatives **227** in moderate to excellent yields, in which the cyclized products are confirmed to be formed from the rearrangement of the normal addition products **184** upon heating (Scheme 5.6) [16].

In 2003, Mizuno et al. reported the cyclopropanation reactions of VDCPs **1** with CHX_3 as the precursor of carbene. It was reported that reactions of diaryl-substituted VDCPs **1** (R^1, R^2 = aryl) with dibromocarbene and dichlorocarbene exclusively gave 1-(dihalomethylene)spiropentanes **228** in moderate to high

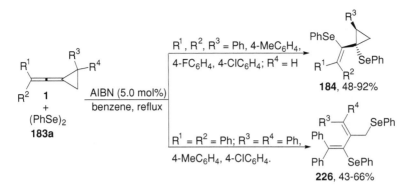

Scheme 5.5 AIBN-mediated reactions of VDCPs **1** with diphenyl diselenide. Reprinted with the permission from Ref. [1]. Copyright 2011 American Chemical Society

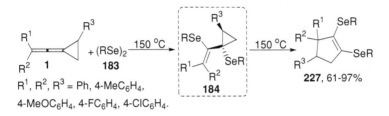

Scheme 5.6 Thermal-induced reactions of VDCPs **1** with diaryl diselenide. Reprinted with the permission from Ref. [1]. Copyright 2011 American Chemical Society

yields, while reactions of monoaryl-substituted VDCPs **1** (R^1 = aryl, R^2 = alkyl) with dihalocarbenes afforded cyclopropylidenecyclopropanes **229** as the major products with the formation of a small amount of **228**. It was also observed that products **229** can be easily converted to products **228** quantitatively in refluxing toluene for 2 h (Scheme 5.7) [17].

5.3 Cycloadditions of VDCPs

Iodobenzene diacetate-mediated reactions of VDCPs **1** with phthalhydrazide can give the corresponding [3 + 2] cycloaddition products **232** in good yields under mild reaction conditions [18]. It was believed that in these reactions, phthalhydrazide was transformed to a 1,3-dipole intermediate in the presence of iodobenzene diacetate. First, iodobenzene diacetate oxidized phthalhydrazide to phthalazine-1,4-dione **230a** [19, 20], which was an equivalent of 1,3-dipole intermediate **231a**. The 1,3-dipole intermediate **231a** reacted with the C1–C1' double bond of VDCPs **1** to give the corresponding cycloaddition products **232** (Scheme 5.8).

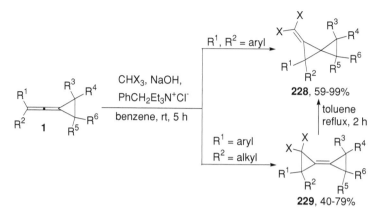

Scheme 5.7 Reactions of VDCPs **1** with dihalocarbenes. Reprinted with the permission from Ref. [1]. Copyright 2011 American Chemical Society

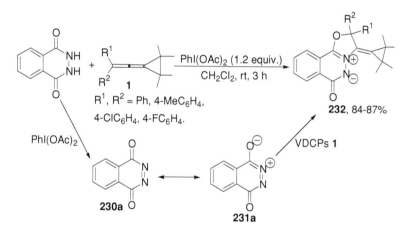

Scheme 5.8 PhI(OAc)$_2$-mediated reactions of VDCPs **1** with phthalhydrazide. Reprinted with the permission from Ref. [1]. Copyright 2011 American Chemical Society

In 2010, Shi et al. reported the 1,3-dipolar cycloaddition reactions of VDCP-diesters **1** with aromatic diazomethanes generated in situ from the corresponding aromatic aldehydes and tosylhydrazine mediated by a base to produce pyrazole derivatives **233** and **233′** in good total yields, with the former as the major, under mild conditions (Scheme 5.9) [21].

On the basis of the above results, plausible mechanism for this [3 + 2] cycloaddition reaction is shown in Scheme 5.10 with the reaction of VDCP **1h** and benzaldehyde as the model. Initially, condensation of tosylhydrazine with benz-aldehyde followed by treatment with a base leads to a solution of benzaldehyde tosylhydrazone salt, which upon heating to 50 °C gives phenyl diazomethane

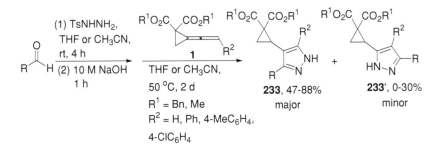

Scheme 5.9   1,3-Dipolar cycloaddition reactions of VDCP-diesters **1**

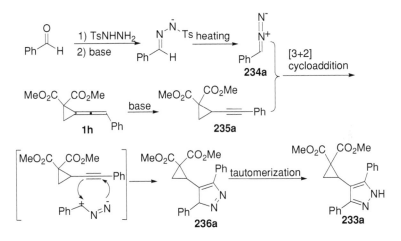

Scheme 5.10   Plausible mechanism for the formation of product **233a**. Reprinted with the permission from Ref. [21]. Copyright 2011 American Chemical Society

234a. Meanwhile, under the basic condition, VDCP **1h** tautomerizes to its alkyne isomer **235a**, which furnishes intermediate **236a** through [3 + 2] cycloaddition with phenyl diazomethane **234a**. The proton transfer in intermediate **236a** affords the final product **233a**.

5.4   Oxidation Reactions of VDCPs

In 1961, Hartzler et al. reported the ozonolysis reaction of VDCP **1a** in ethanol, in which cyclopropyl hydroxylester **237** was formed in good yield (Scheme 5.11) [1, 22]. Crandall et al. examined this transformation in detail and it was reported that the addition of VDCP **1a** to a saturated solution of ozone (1.0 equiv.) in CDCl$_3$ at −61 °C gave acetone, 3-phenylcyclobutane-1,2-dione and phenylsuccinic

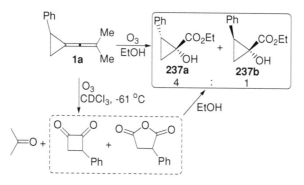

Scheme 5.11 Oxidative reaction of VDCP **1a** with O₃. Reprinted with the permission from Ref. [1]. Copyright 2011 American Chemical Society

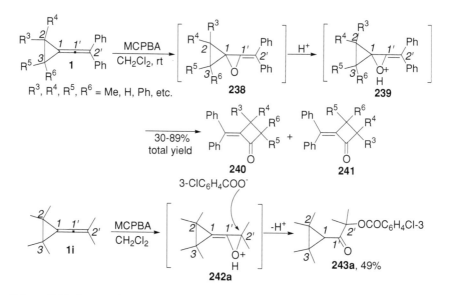

Scheme 5.12 Oxygenation of VDCPs **1** with MCPBA. Reprinted with the permission from Ref. [1]. Copyright 2011 American Chemical Society

anhydride. Treatment of the latter two compounds with excess ethanol can give the esters **237**, which clearly indicates that these two compounds are precursors to the esters (Scheme 5.11) [23].

In 1992, Mizuno et al. reported the regioselective MCPBA-oxidation reactions of VDCPs **1** and the results indicate that the regioselectivity in the epoxidation reactions strongly depends on the substituents on C2′ position of VDCPs **1**. When R^1 and R^2 are both phenyl groups, 2-methylene-cyclobutan-1-ones **240** and **241** are formed singly; when VDCP **1i**, in which R^1 and R^2 are both methyl groups, was used as the substrate, cyclopropyl keto ester derivative **243a** is obtained

Scheme 5.13 Oxygenation of VDCPs **1** with dimethyldioxirane. Reprinted with the permission from Ref. [1]. Copyright 2011 American Chemical Society

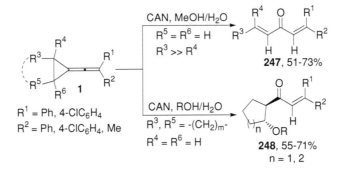

Scheme 5.14 CAN-mediated oxidative rearrangement reactions of VDCPs **1**. Reprinted with the permission from Ref. [1]. Copyright 2011 American Chemical Society

exclusively (Scheme 5.12) [24, 25]. Proposed mechanisms for these transformations are also illustrated in Scheme 5.12. Differences between these two oxidation reactions may be ascribed to the steric and electronic effects of the phenyl groups in C2′ position of VDCPs **1**: it is worthy of noting that phenyl groups at the C2′ position cannot be coplanar with the C1′–C2′ plane for a steric reason.

Oxygenation of VDCPs **1** with dimethyldioxirane in CH_2Cl_2 in the presence of 18-crown-6 gave methylenecyclopropanes **244**. Further oxygenation of **244** with dimethyldioxirane gave cyclobutanones **245** and **246** (Scheme 5.13).

In 2008, Huang et al. reported cerium(IV) ammonium nitrate (CAN)-mediated oxidative rearrangement reactions of VDCPs **1**, which resulted in unsymmetrical divinyl ketones **247** and enone derivatives **248** in moderate to good yields with excellent regio- and stereo-selectivities (Scheme 5.14) [26].

Plausible mechanism for these transformations is shown in Scheme 5.15. VDCPs **1**, in the presence of Ce(IV), undergo oxidative electron transfer to afford cationic radical **249**. For a review, see [27]. The following nucleophilic attack of the solvent as MeOH at the cyclopropyl ring may cause the rearrangement to produce the ring-opened radical intermediate **250** [28]. Intermediate **250** can be further oxidized by another molecule of CAN to give cation **251**, which is quenched by water in the solvent to produce intermediate **252**. Subsequent enol rearrangement of the corresponding intermediate **252** affords product **248** or

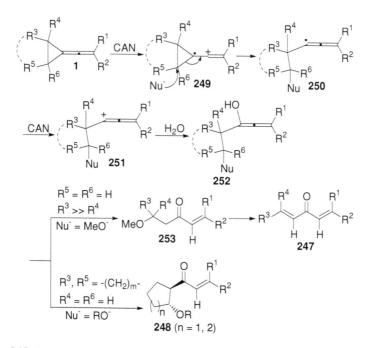

Scheme 5.15 Plausible mechanism for the CAN-mediated oxidative rearrangement reactions of VDCPs **1**. Reprinted with the permission from Ref. [1]. Copyright 2011 American Chemical Society

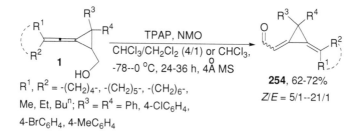

Scheme 5.16 Oxidative isomerization of VDCPs **1** in the presence of TPAP and NMO

intermediate **253**. Further elimination of a molecule of MeOH from **253** furnishes product **247**. While as for **248**, the *trans*-isomer in a sterically hindered ring does not progress to the corresponding divinyl ketones.

In early 2011, Shi et al. reported that in the presence of tetrapropylammonium perruthenate (TPAP) and 4-methylmorpholine *N*-oxide (NMO), the oxidative isomerization reactions of VDCPs **1** can be achieved to afford dimethylenecyclopropane aldehydes **254** in moderate yields (Scheme 5.16) [29].

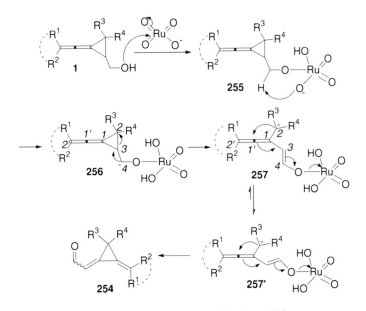

Scheme 5.17 Plausible mechanism for the formation of products **254**

Plausible mechanism for the formation of products **254** is outlined below: first, VDCPs **1** reacts with tetrahedral perruthenate ion $Ru^{VII}O_4^-$ to give intermediate **255** [30–32]. After a quick hydrogen migration, the cyclopropyl ring opening of intermediate **256** takes place to afford intermediates **257** and **257'**, which undergoes recyclization to give products **254** (Scheme 5.17). It is plausible that the C3–C4 double bond in intermediate **257** interacts with the sterically large R^2 group, so intermediate **257'** is more stable than intermediate **257**. Therefore, intermediate **257** is assumed to be transformed into **257'** during the reaction, leading to the formation of Z-**254** as the major isomer. The electronic effect of the substituents R^3 and R^4 groups may also have some influence on the Z/E geometric selectivity.

References

1. Shi M, Shao LX, Lu JM, Wei Y, Mizuno K, Maeda H (2010) Chemistry of vinylidenecyclopropanes. Chem Rev 110:5883–5913
2. Shao LX, Zhang YP, Qi MH, Shi M (2007) Lewis acid catalyzed rearrangement of vinylcyclopropenes for the construction of naphthalene and indene skeletons. Org Lett 9:117–120 (and references therein)
3. Lu JM, Shi M (2008) LDA-mediated selective addition reaction of vinylidenecyclopropanes with aldehydes, ketones, and enones: facile synthesis of vinylcyclopropenes, allenols, and 1,3-enynes. Org Lett 10:1943–1946
4. Lu JM, Shi M (2009) Lithium diisopropylamide-mediated carbolithiation reactions of vinylidenecyclopropanes and further transformations of the adducts. Chem Eur J 15:6065–6073

5. Huang JW, Shi M (2005) Carbolithiation of *gem*-aryl disubstituted methylenecyclopropanes. Org Biomol Chem 3:399–400
6. Paradies J, Erker G, Fröhlich R (2006) Functional-group chemistry of organolithium compounds: photochemical [2 + 2] cycloaddition of alkenyl-substituted lithium cyclopentadienides. Angew Chem Int Ed 45:3079–3082
7. Miller CJ, O'Hare D (2005) A new phosphine-functionalised [1]ferrocenophane and its use in the functionalisation of mesoporous silicas. J Mater Chem 15:5070–5080
8. Chou PK, Dahlke GD, Kass SR (1993) Unimolecular rearrangements of carbanions in the gas phase. 2. Cyclopropyl anions. J Am Chem Soc 115:315–324
9. Creary X (1977) 3,3-Dimethylallenyllithium. Reaction with electrophiles leading to carbenoid, electron transfer, and nucleophilic processes. J Am Chem Soc 99:7632–7639
10. Moreau JL (1980) In: Patai S (ed) The chemistry of ketenes, allenes and related compounds. Wiley, New York, p 363
11. Huynh C, Linstrumelle G (1983) Prop-2-ynyl- and propadienyllithium reagents. Regiocontrolled synthesis of allenic compounds. J Chem Soc Chem Commun 1133–1134
12. Faza ON, López CS, de Lera AR (2007) Sulfoxide-induced stereoselection in [1,5]-sigmatropic hydrogen shifts of vinylallenes. A computational study. J Org Chem 72:2617–2624
13. Lu BL, Lu JM, Shi M (2009) Butyl lithium (^{n}BuLi)-mediated carboxylation of vinylidenecyclopropanes with CO_2. Tetrahedron 65:9328–9335
14. Lu BL, Lu JM, Shi M (2010) LDA-mediated domino carbolithiation reactions of vinylidenecyclopropanes with but-3-yn-2-one and 1-phenylprop-2-yn-1-one. Tetrahedron Lett 51:321–324
15. Shi M, Lu JM (2006) Reactions of vinylidenecyclopropanes with diphenyl diselenide in the presence of AIBN and further transformation to produce new naphthalene derivatives. J Org Chem 71:1920–1923
16. Shi M, Lu JM, Xu GC (2005) Ring-opening reactions of arylvinylidenecyclopropanes with diaryl diselenide upon heating: formation of 1,2-diarylselenocyclopentene derivatives. Tetrahedron Lett 46:4745–4748 (and references therein)
17. Maeda H, Hirai T, Sugimoto A, Mizuno K (2003) Cyclopropanation of vinylidenecyclopropanes. Synthesis of 1-(dihalomethylene)spiropentanes. J Org Chem 68:7700–7706
18. Liu LP, Lu JM, Shi M (2007) PhI(OAc)$_2$-mediated novel 1,3-dipolar cycloaddition of methylenecyclopropanes (MCPs), vinylidenecyclopropanes (VCPs), and methylenecyclobutane (MCB) with phthalhydrazide. Org Lett 9:1303–1306
19. Landa A, Seoane C, Soto JL (1974) Nitrogen heteroconjugations. VI. Reactions with diazaquinones. An Quim 70:962–965
20. Lora-Tamayo NP, Pardo M, Soto JL (1975) Azapolycyclic compounds. X. Oxidation of diazaquinone adducts. An Quim 71:400–405
21. Wu L, Shi M (2010) 1,3-Dipolar cycloaddition reactions of vinylidenecyclopropane-diesters with aromatic diazomethanes generated in situ. J Org Chem 75:2296–2301
22. Hartzler HD (1961) Carbenes from derivatives of ethynylcarbinols. The synthesis of alkenylidenecyclopropanes. J Am Chem Soc 83:4990–4996
23. Crandall JK, Schuster T (1990) The ozonolysis of alkenylidenecyclopropanes. J Org Chem 55:1973–1975
24. Sugita H, Mizuno K, Saito T, Isagawa K, Otsuji Y (1992) Regioselective MCPBA-oxidation of alkenylidenecyclopropanes: a convenient synthesis of 2-methylenecyclobutan-1-ones and cyclopropyl keto esters. Tetrahedron Lett 33:2539–2542
25. Crombie L, Maddocks PJ, Pattenden G (1978) Monoterpene synthesis via alkenylidenecyclopropanes: metal reductions and peracid oxidations. Tetrahedron Lett 19:3483–3486
26. Su CL, Huang X, Liu QY (2008) CAN-mediated highly regio- and stereoselective oxidation of vinylidenecyclopropanes: a novel method for the synthesis of unsymmetrical divinyl ketone and functional enone derivatives. J Org Chem 73:6421–6424

27. Nair V, Balagopal L, Rajan R, Mathew J (2004) Recent advances in synthetic transformations mediated by cerium(IV) ammonium nitrate. Acc Chem Res 37:21–30
28. Siriwardana AI, Nakamura I, Yamamoto Y (2003) Addition of hydrogen halides to alkylidenecyclopropanes: a highly efficient and stereoselective method for the preparation of homoallylic halides. Tetrahedron Lett 44:985–987
29. Lu BL, Shi M (2011) Oxidative isomerization of vinylidenecyclopropanes to dimethylenecyclopropanes and Brønsted acid-catalyzed further transformation. Eur J Org Chem 243–248
30. Lee DG, Wang Z, Chandler WD (1992) Autocatalysis during the reduction of tetra-n-propylammonium perruthenate by 2-propanol. J Org Chem 57:3276–3277
31. Roček J, Ng CS (1974) Chromium(IV) oxidation of aliphatic aldehydes. J Am Chem Soc 96:1522–1529
32. Roček J, Ng CS (1973) The role of hydrate formation in the chromium(VI) oxidation of aldehydes. J Org Chem 38:3348–3350

Chapter 6
Concluding Remarks and Perspectives of VDCPs

During the past years, the chemistry of VDCPs has rapidly developed. VDCPs, especially those with aromatic group(s) on the allene and/or cyclopropyl ring moiety, show diverse reaction patterns in the presence of Lewis acids, Brønsted acids, transition metal catalysts, etc. The chemistry of VDCPs greatly depends on the substituents on the allene and cyclopropyl ring moieties. Lewis or Brønsted acid-mediated chemistry of VDCPs has aroused a renaissance of cationic intermediates. Novel transformations of VDCPs developed during the past years have resulted in the synthesis of aromatic compounds, heterocyclic compounds, and some other useful products. It is believed that with continued investigations in this area, many new reactions and more useful chemistry of VDCPs will be found in the near future.

L. Shao et al., *Chemical Transformations of Vinylidenecyclopropanes*,
SpringerBriefs in Molecular Science, DOI: 10.1007/978-3-642-27573-9_6,
© The Author(s) 2012